THE POLITICS OF KNOWLEDGE
IN CENTRAL ASIA

There has been great interest and investment in reforming scientific institutions throughout the post-Soviet world. To date however, no thorough analysis of the role of organized intellectual activity in the region is available. Through careful historical and ethnographic research and the extensive use of local scholarly works, this book provides a persuasive and careful analysis of the production of knowledge in Central Asia. The author demonstrates that classical theories of scientific revolution or science and society are inadequate for understanding the science project in Central Asia. Instead, a critical understanding of local science is more appropriate. In the region, the professional and political ethos of Marxism–Leninism was incorporated into the logic of science on the periphery of the Soviet empire. Local academics assimilated, negotiated and resisted its priorities in their work. Similarly, after the end of Soviet rule, they interacted with the new post-Soviet 'logic of the market' and democratic ethos of science in an effort to refashion a new science for their new society. The scientists' work to establish themselves 'between Marx and the market' is therefore creating new political economies of knowledge at the periphery of the scientific world system.

Sarah Amsler is Senior Lecturer in Sociology at Kingston University, UK. She held previous posts at the American University–Central Asia in Kyrgyzstan and Central Asian Resource Centre in Kazakhstan. Her fields of expertise are the critical sociology of knowledge, the sociology of cultural institutions and the politics of science, particularly in postcolonial and post-Soviet societies.

CENTRAL ASIA RESEARCH FORUM
Series Editor: Shirin Akiner
School of Oriental and African Studies
University of London

Other titles in the series:

SUSTAINABLE DEVELOPMENT
IN CENTRAL ASIA
Edited by Shirin Akiner, Sander
Tideman and John Hay

QAIDU AND THE RISE OF THE
INDEPENDENT MONGOL STATE
IN CENTRAL ASIA
Michal Biran

TAJIKISTAN
Edited by Mohammad-Reza Djalili,
Frederic Gare and Shirin Akiner

UZBEKISTAN ON THE
THRESHOLD OF THE TWENTY-
FIRST CENTURY
Tradition and survival
Islam Karimov

TRADITION AND SOCIETY
IN TURKMENISTAN
Gender, oral culture and song
Carole Blackwell

LIFE OF ALIMQUL
A native chronicle of nineteenth
century Central Asia
Edited and translated by
Timur Beisembiev

CENTRAL ASIA
Aspects of transition
Edited by Tom Everrett-Heath

THE HEART OF ASIA
A history of Russian Turkestan
and the Central Asian Khanates
from the earliest times
Frances Henry Skrine and
Edward Denison Ross

THE CASPIAN
Politics, energy and security
Edited by Shirin Akiner and Anne
Aldis

ISLAM AND COLONIALISM
Western perspectives on Soviet Asia
Will Myer

AZERI WOMEN IN TRANSITION
Women in Soviet and post-Soviet
Azerbaijan
Farideh Heyat

THE POST-SOVIET DECLINE
OF CENTRAL ASIA
Sustainable development and
comprehensive capital
Eric Sievers

THE POLITICS OF KNOWLEDGE IN CENTRAL ASIA

Science between Marx and the market

Sarah Amsler

Routledge
Taylor & Francis Group

LONDON AND NEW YORK

First published 2007
by Routledge
2 Park Square, Milton Park, Abingdon, Oxon OX14 4RN

Simultaneously published in the USA and Canada
by Routledge
270 Madison Ave, New York, NY 10016

*Routledge is an imprint of the Taylor & Francis Group,
an informa business*

© 2007 Sarah Amsler

Typeset in Times New Roman by
Newgen Imaging Systems (P) Ltd, Chennai, India
Printed and bound in Great Britain by
TJI Digital, Padstow, Cornwall

British Library Cataloguing in Publication Data
A catalogue record for this book is available
from the British Library

Library of Congress Cataloging in Publication Data
A catalog record for this book has been requested

ISBN10: 0–415–41334–6 (hbk)
ISBN10: 0–203–96174–9 (ebk)

ISBN13: 978–0–415–41334–3 (hbk)
ISBN13: 978–0–203–96174–2 (ebk)

CONTENTS

PREFACE

I began researching this book at a historical moment, both auspicious and tragic, seven years after the final disintegration of the Soviet Union, while working as a sociologist in Central Asia during the late 1990s. The nature of my early encounters with Central Asia and the circumstances under which I have since worked in the region have decisively shaped my approach to the sociological study of organized knowledge in these societies. I, like many other foreign social scientists working in the former Soviet Union during this period, was initially recruited by a 'development' organization – in my case the Open Society Institute, which at the time sponsored an academic fellowship programme called the Civic Education Project.[1] As I settled awkwardly into a position as assistant chair of the new Sociology Department at the American University–Central Asia, I began to wonder why the organization had recruited me, a newly qualified American academic who knew little about the society in which she was hired to teach, and also what epistemological, political and cultural prejudices had allowed me to accept a position whose authority was clearly predicated on Occidentalist generalizations about the universality, superiority and progressive nature of 'Western' knowledge. The sociology and politics of knowledge, and particularly as regards colonialism and postsocialism, were implicit in this work from the start.

Within the post-Soviet academy, all that was once solid had indeed melted into air and the future seemed both wide open and frighteningly impossible. The authority of academic knowledge, particularly in the social sciences, was heavily damaged by its politicization within Soviet society. It was being reclaimed through new associations with the very ideologies of positivism and empiricism that had epitomized its antithesis, bourgeois 'pseudoscience', only a decade before. While there was ostensibly expanded space for intellectual experimentation, imagining what this might entail was difficult. The reconstruction of existing boundaries of legitimate knowledge was thus experienced as a crisis as much as an opportunity. Beyond the practical challenges of post-Soviet higher education, there were also palpable currents of wider contention: discourses on the inferiority of 'local' or 'Soviet' knowledge and the superiority of 'Western' knowledge, fatalistic determinism in sociological theory and research, and polarized responses of unproblematic attraction or irredentist reaction to post-Marxist models of progress in neoliberal development agendas.

In this environment, my interest in theorizing social scientific 'knowledge' met with both enthusiasm and resistance. There was not only little time for luxurious meta-theorizing – empirical research needed to be conducted and disseminated, courses needed to be designed and taught, prolific institutional regulations needed to be satisfied – but some academics also felt that it was intellectually unnecessary. Problems encountered were taken for granted as natural elements of 'the transition', a slippery and ideologically laden concept that we inherited from an earlier epoch of postcolonial modernization projects and reinterpreted to suit the post-Soviet one. Historical process had spoken: 'Marx' (and much of what this name entailed) and the 'East' had lost, and the 'market' and the 'West' had won. What would be the purpose of looking backward for answers, into a history characterized by what were publicly characterized as humiliating errors of intellectual and professional judgement? Or of peering inward, when decades of cultural work and workers seemed so obviously proved to be inadequate for understanding or creating the 'good society'? And what was the point of discussing the current politics of knowledge when it was now clear and possible that true knowledge could *not* be 'political'; when the very possibility of the 'sociology of knowledge' detracted from the little legitimacy that social scientists had established under the Soviet regime?

There were also more political reservations. For example, attempts to democratize the research process; include interlocutors as partners instead of objectifying them as 'subjects'; and inviting participants to comment on tentative findings were greeted with ambivalence. Why would I deliberately discredit myself by trying to minimize the authority bestowed upon me as a foreign 'expert'? In many respects, the dominant academic culture in the region respects hierarchy and deference, expertise and pretences to neutral objectivity; in short, it rewards as 'science' much that I deliberately call into question. Some suspected that the 'interactive' approach to research was a manipulative and paternalistic experiment similar to professional 'trainings' now so often organized by international educational organizations.[2] A few were angry about the insinuation, however benign I initially imagined it to be, that their knowledge was in any way 'political' and resisted being interpreted as political as well as intellectual actors. Finally, some distrust foreign researchers, whom they fear – not entirely without cause – will steal their ideas and slander their reputations in foreign-language journals that they can neither access nor read. In short, through this work I became aware that epistemological and political architectures, even of methods that aim to dismantle power relations, are structurally embedded within these very relations.

As I began to explore this phenomenon more systematically, I had another – this time textual – encounter which reoriented the project. I began to accumulate books and articles, which I felt described and in some cases theorized the state of social science in Central Asia. However, they were written not about post-Soviet societies, but by and about social scientists working elsewhere at other times in the postcolonial world: India in the 1960s, Latin America in the 1970s and Africa in the 1980s. My analytical aperture widened: problems of organized knowledge production in Central Asia cannot be adequately understood if we view them only through the narrow national and regional lenses through which we had for so long been accustomed to looking. They beg questions about knowledge, power and capital at the global level; about the relationship between the colonial and Soviet experiences, on the one hand, and neocolonialism and post-Soviet independence on the other; about the relationship between 'globalization' and 'postsocialism'. It occurred to me that social science in Central Asia may not, as it were, be undergoing a simple 'transition' from ideology to truth, as often assumed by narratives of de-Sovietization and development. This book instead paints a different picture of a complex social and cultural institution that has been continuously re-imagined as part of shifting encounters between the logics of science and power, and of late, the 'market'.

Reflections on power and knowledge in the field

The book has also been shaped, less intentionally if not less reflexively, by the usual ethnographic suspects: culture shock, language barriers, problematic access to people and documents, strained rapport with interlocutors, role conflicts in the field, and uneven power relations between the researcher and the researched. Other researchers of former Soviet countries argue that there is, in addition, 'something peculiarly postsocialist about the inevitable complexity of fieldwork relations' in these societies. They cite the impact of Cold-War ideologies on mutual impressions of researcher and researched, the as-yet-untheorized differences of everyday social organization in non-capitalist cultures, the way that people in these formerly closed societies interpret the intrusion of foreign observers, and the ambiguous relationship between detachment and engagement in the post-Soviet field (Dudwick and De Soto 2000). However, if we are to make any sense out of this shared experience, it is necessary to move beyond its recognition and theorize how the particular features of postsocialist ethnography are related to broader issues of power and knowledge embedded in the imperial politics of the academy in the region.

First and foremost is the problem of how to negotiate, if possible to deconstruct, the Orientalist, Occidentalist and colonial subtexts of social research in Central Asian societies. Some have framed this problem as a post-Cold-War clash between 'triumphant' capitalist researchers and disappointed and 'defeated' Soviet citizens (Liu 2003; Zanca 2000: 153). I suggest that it is also linked to institutionalized structures of power and domination within Central Asian society itself, many of which have been obscured by well-intentioned but misguided 'post-power' discourses of globalization and civil society in recent years.[3] The people of Central Asia are self-consciously observed and evaluated, and are therefore often wary of the motives and intentions of foreign researchers. This is particularly true for social elites, including many of the academics and intellectuals discussed in this research, for whom national independence wrought not only professional dislocation but also severe losses of economic privilege, social power and cultural prestige. The notion that all scholars were 'liberated' from the very social structures in which they were gaining status during the 1980s is a bitter irony for those who were invested, both professionally and intellectually, in the institutions of Soviet science. Since independence, asymmetrical power relations between foreign and indigenous researchers have been exacerbated by the emergence of new inequalities, such as age, political orientation and access to English-language or American and European education, which compound existing hierarchies of ethnicity, gender, region and party affiliation. In some cases, scholars' work and professional identities have been simplistically branded as naïve, illegitimate and ideological – to use the word in its contemporary pejorative sense, 'Soviet'. Academics who once saw themselves as architects or administrators of a formidable empire, once the guardians of the truth about social reality, have become the exotically observed and passé. This has obvious implications for research relations, which are therefore also experienced as political encounters.

Second, this research has been an exercise in comprehending and translating theoretical dissonances that were revealed between my interlocutors and my self. The most vivid example of such 'talking past' actually comes from a colleague, however, who was once accused of denying the existence of the Kyrgyz nation after presenting a conference paper on the social construction of ethnicity in Kyrgyzstan. It emerged that a statement which would not even raise eyebrows in a setting of shared epistemological *doxa* could easily require hours, days or perhaps even years of preliminary discussion in a more heterogeneous environment. As other ethnographers of Central Asian societies have argued, the success of the interpretive endeavour depends not only on how well one can master the 'epistemic negotiations' that are vital for cross-cultural understanding, but also on how well the analyst comprehends the larger social and political contexts that ground the epistemologies, and how well she 'answers not for the impartiality or replicability of her research, but for the situated knowledge she has collaborated with her informants to produce' (Adams 1999: 331).[4] This has been a particular challenge, as it is precisely this sort of knowledge that is the main focus of this book.

Finally, as a sociologist studying sociologists with whom I also worked, this research raised questions about how to negotiate 'objectivity' and 'engagement'. Ultimately, I never resigned myself to the advice of a trusted friend, a young Kazakh professor, who advised me to enter into power relations or face exclusion from the academic community. 'Be instrumental', he said, 'use your power. That's how it works here'. In many senses, he was right. That, unfortunately, is rather 'how it works' there at the moment; power relations are an integral part of academic practice in Central Asia, as they are elsewhere. However, I decided, perhaps against the rules of 'good' anthropology where one strives to conform for 'rapport' and 'acceptance', that I wished not to be bounded by this fatalistic essentialism, and instead attempted to preserve a methodological faith in the possibility of democratizing the research process, even in this imbalanced context. As with all decisions in social research, this choice closed certain doors and opened others, including to relationships with people and ideas that have been excluded from the traditional structures of academic discourse but who are playing major roles in the transformation of the social sciences in Central Asia.

I have paused on these methodological issues because they are central to the way in which I articulate the relationship between theory, method and practice in this book. In addition, they contextualize the research process within some of the political, cultural and economic forces that both inspired and constrained it.[5] These points are therefore understood as integral to the research, rather than as auxiliary concerns.

The critical stance: some premises

I have made every attempt to construct a valid representation of the sociology and politics of knowledge in Central Asia, one which accounts for both the 'logic of science' and the 'logic of practice' (Bourdieu 1992) – which would, in other words, be loyal both to the principles of critical sociological theory and to the subjective meanings and values that Central Asian sociologists attach to their own work. Another story, guided by other experiences, political proclivities and theoretical orientations, would be different. I am not a detached observer of Central Asian society. In addition to a theoretical interest in the sociology of knowledge, I have political and moral concerns about the democratization of knowledge and about the social consequences of the politics of truth in the region.

The interpretive work in this book therefore draws on two sociological traditions that may be described as 'critical'. It is first informed by theories of knowledge that question grand narratives of modern scientific progress and aim to expose the political, economic and cultural foundations – and where appropriate the human consequences – of what might be called the underside of enlightenment. Critical theories of knowledge, which were elaborated in relation to pre-fascist Europe and became ascendant throughout Western Europe and the United States during the 1960s and 1970s, assume renewed significance in Central Asia. Here, 'scientific knowledge' is portrayed as a redemptive political force, and it has become almost heretical to challenge the origins or consequences of its claims to epistemological authority or to interrogate the cultural meaning of its intellectual products. Further, academic knowledge is linked to technocratic action, underpinned by instrumental rationality and driven by an unexamined belief in the promises of rational scientific progress (Torres 1999). These underlying assumptions lead researchers into teleological studies which may explore the causes and effects of social change but neglect to consider the nature and politics of this change itself, for 'when the prophetic and the progressive are important to social life, their inscription in social and educational sciences is an orthodoxy that makes it difficult for us to perceive them as effects of power' (Popkewitz 1991: 27). Critical theory reminds us that in a society where social scientists are considered to be physicians who diagnose and cure the ills of a sick society, it is also crucial to investigate the limitations and possibilities of the 'healers' themselves.

The second 'critical' dimension of this research is its orientation towards humanist sociology. In the tradition of Max Weber, C. Wright Mills and others it defines sociological research as a moral responsibility as well as an intellectual endeavour, believing that in questions of human freedom, 'nothing is less innocent than non-interference' (Bourdieu 1999: 629). My analysis of academic knowledge and culture thus inevitably differs from that of policy makers, aid workers and anthropologists who have also written on similar themes. Its provenance in praxis has made it methodologically challenging, and I hope that my attempts to accommodate both symmetrical analytics and normative practice, ethnography and critical theory, and observation and engagement will raise provocative questions about the sociology of knowledge in Central Asia and other places subject to both nationalizing and neo-colonial 'development' in the conditions of postsocialist global capitalism.

ACKNOWLEDGEMENTS

Many people have contributed to this book in various ways; particular thanks go to Jens Binder, Todd Drummond, Nancy Hanrahan, Britta Korth, Baktygul Kulusheva, Ludmilla Liubimova, Martha Merrill, Kimberly Montgomery, Stefanie Ortmann, Farida Osmonova, Madeleine Reeves, Margaret Reeves, Ainoura Sagynbaeva, Balihar Sanghera, Richard Sennett, Burul Usmanalieva, Nienke van der Heide, Elizabeth Weinberg and Tatiana Yarkova. At the institutional level I must thank ACCELS, members of the Faculty of Socio-Political Sciences at the Bishkek Humanities University, the departments of Sociology and International Politics at the American University–Central Asia, and the Central Asia Resource Center in Kazakhstan. This and all my work is made possible through the intellectual challenge, emotional fortitude, unconditional love and support, and domestic camaraderie of my husband, Mahmood Delkhasteh. Although this book obviously emerged from extensive collaboration, I alone am responsible for its content.

INTRODUCTION

How did it happen that we so quickly 'forgot' about the decades-long preaching of communist ideology, that we believed in as the 'sole truth' and 'sole science'? Is it proper that, not having clarified these painful and core questions for ourselves, we have begun to elaborate a 'new ideology' as if the former one did not exist, as if those people who now so energetically took the ideology of 'national rebirth' or, let's say, the ideology of the 'all-consuming market' did not also militantly struggle for the realisation of 'communist ideas'?

(Asanova 1995)

How is it possible for man to continue to think and live in a time when the problems of ideology and utopia are being radically raised and thought through in all their implications?

(Mannheim 1936: 42)

Despite great interest and investment in reforming cultural institutions throughout the post-Soviet world, there has been little rigorous research into either this project or the more specific dynamics of organized intellectual activity in the region. Sociologists of knowledge and science who in the past have mobilized en masse to analyse lesser upheavals in scientific and intellectual life have remained curiously silent about the fate of ideas in post-Soviet societies. This is particularly true in what were once the 'borderland republics' of Central Asia, where academic and scientific institutions are often treated as development projects or anthropological exercises as opposed to subjects of legitimate theoretical analysis. This can be explained in a number of ways: a historical apartheid between Cold-War-era 'area studies' and mainstream sociological theory, the marginalization of Soviet academics and intellectuals within the global science system, their preoccupation with the daily hardships of academic practice, and Orientalist ideologies and prejudices which led scholars to conclude that there was no legitimate science under the Soviet regime, that no serious work was produced during that time, and therefore there was nothing worth consideration.[1]

Whatever the reason, this oversight has impoverished the understandings of the complex politics of organized knowledge in the region and its potential role in social change and human freedom. It also compromises our understanding of the sociology of knowledge and science more generally, particularly as regards intellectual activity in marginalized academic communities and under new 'post-communist' conditions of neoliberal hegemony in former Soviet space (Outhwaite and Ray 2005). The continuing intersection of knowledge and power within the academy means that knowledge reform must not be taken for granted as 'inherently progressive and truth-producing', as it is often interpreted in reformist discourse (Beliaev and Butorin 1982; Popkewitz 1991). As the Central Asian writer Karybek Baibosunov (1993) has noted of the social sciences, knowledge fields 'are enduring major cataclysms [as they are] freed from an ideological path and seek to raise influence on new trends'. It is also necessary to understand how these changes are influencing the production of organized knowledge itself, and to critically evaluate the new ideological paths which have replaced the old and which are shaping contemporary intellectual practices.

This book is written as a contribution to this conversation and can be read on two levels. Most broadly, it is an exploration into the cultural meaning and consequences of the modern 'science project' in the non-Russian republics of the former Soviet Union, the effect of the Soviet regime and its collapse on efforts to institutionalize academic social science, the impact of these efforts on shaping the knowledge about the 'post-Soviet', the necessity of non-Western perspectives on the 'globalization' of knowledge production, and the shifting relationship between knowledge and power in the academy under politico-economic regimes of both state socialism and late capitalism, or what might also be referred to as the comparative political economy of truth.

The exploration of these broad themes is grounded in the sociological and historical study of the institutionalization of a single social scientific field, sociology, in Kyrgyzstan (formerly the Kirgiz Soviet Socialist Republic). It integrates historical research, ethnography, interviews and content analyses of academic and popular publications on social science with critical theories of knowledge to illustrate how conceptions of 'science' and 'truth' are historically contingent and have been negotiated through professional practices within the Central Asian academy. This focus allows for a deep understanding of the diversity of everyday practices of knowledge production in post-Soviet space. Despite considerable attention to 'national' and 'international' development, local institutional contexts are crucially important in shaping knowledge outcomes and experiences. Comparative research on 'indigenous' or 'national' sociologies provides excellent insight into the political economy of postcolonial science, as will be discussed in the following chapter, but tells us little about how social forces are engaged by academics themselves. Similarly, studies of a general 'Soviet' sociology are instructive but often fail to appreciate the heterogeneity of the Soviet experience, especially in cultural life. Although important to understand the dynamics of power and knowledge at the macro level, it is therefore also vital to examine how local power structures and cultural practices mediate these forces.

The social scientific community within Central Asia lends itself to this analysis, for here, a modernist ideal of scientific politics has long existed side by side and in permanent tension with deep scepticism about the politicization of knowledge itself. This dichotomy gives rise to a series of fascinating professional projects which span the Soviet and post-Soviet experiences and which are very specifically related to the articulation of boundaries between 'knowledge' and 'power'. Since the mid-twentieth century, social scientists have worked continuously to align the relationship between social science and politics in order to transform a heteronymous field of knowledge production and scientific practice into an autonomous one, or from a field whose development is dominated by external forces to one which is self-producing and reproducing and which can exert influence in socio-political practice. Although varying across geopolitical contexts – internal colonialism, national independence and capitalist dependency – projects to institutionalize and reform sociology have all been grounded in a need to define and articulate relationships between 'truth' and 'power' which enable social scientists to negotiate their professional identity in ways that are commensurate with the logics of both science and politics. While analysed in a post-Soviet context, this can also be understood as a more general tension between establishing 'sociological relevance' and 'social relevance', or between criteria used to measure the intellectual validity of social scientific knowledge and those used to evaluate its societal significance, which has been a consistent tension in social science, particularly in postcolonial societies (Joshi 1995: 82).

This book will therefore focus especially on the 'boundary-work' done by Kyrgyzstani academics to define the field of sociology during the late socialist period (1985–91) and in the decade following national independence (1991–2001). This is an alternative to the traditional approaches of the 'institutionalization' or 'development' of knowledge fields and professions. Boundary-work, or the 'rhetorical strategy of promoting particular ideologies of science' (Gieryn 1983), is an analytical concept used by sociologists of knowledge to the means by which fields of legitimate knowledge are constructed, maintained, transformed and broken down, both within scientific communities (Camic and Xie 1994; Fuchs 1986; Gieryn 1983; Kuklick 1980) and in the public sphere (Fisher 1990; Gieryn et al. 1985). It builds on the theory that disciplines are socially constructed as opposed to naturally occurring, but extends this by exploring how and under what conditions they are formed and legitimized, by whom and with what intentions, and how the definition of 'truth' is conditioned by the social and material relationships in which these processes of validation are embedded. The central assumption underlying the concept is that the borders of knowledge units (e.g. the definition of knowledge, its distinction from non-science and pseudoscience, the relationship between knowledge and power, etc.) are not fixed or universal, but rather fluid and negotiated in contests for professional legitimacy, cultural authority and material or social resources (Gieryn 1983).

3

Analysing boundary-work in social science is therefore a way of understanding how and why knowledge is actually produced with a focus on the localized actors and institutions that have an interest in this process (Mulkay 1991).

In this context, social science cannot be seen merely as a site of 'transition', but may also be one of revolution, inertia, struggle, retrenchment or resistance. It grounds practical debates about how to reorganize the intellectual architecture of interrupted worlds and are spaces where alternative ways of knowing about these worlds may potentially be introduced, adopted, challenged and negotiated. The structure and meaning of knowledge are changing here, but not in entirely predictable or systematic ways. Questions about what exists in society, or what may be known to exist, how this might be legitimately ascertained and verified, who has the right to know and speak of knowledge and organize its production, and what role authoritative knowledge can and should have in social life – in other words, epistemology – are debated privately, publicly and in earnest. The production of social knowledge in postsocialist space is often a cultural politics, raw with emotion and linked to power, associated not only with questions of 'national development' or 'globalization' (as it often discursively is) but also to the survival of academic careers and to human emotions of anomie, dilemma and hope in the face of an uncertain future. The intimate relationship between knowledge and power is thus simultaneously assumed, denied and contested in Central Asian social science, and projects to reform knowledge institutions in the region, particularly within the academy, cannot be adequately understood outside this context.

The ethnographic and historical data discussed here challenge dominant perceptions that Soviet science was wholly ideological and post-Soviet science unproblematically 'autonomous'. We will see how a professional ethos of 'Marxism–Leninism' was variously defined and incorporated into the logic of science on the periphery of the Soviet empire, and how academics assimilated, negotiated and resisted its priorities in their work in the pursuit of both social truth and professional prestige.[2] We will see how they later struggled to integrate these practices and epistemologies with the post-Soviet 'logic of the market' and democratic ethos of science in an effort to refashion a new science for their new society, and to construct discourses of truth to facilitate its legitimization in the face of new authoritarianisms. We will also learn why political ideologies were once considered good science and why positivism has become good politics, and how the boundaries of science have shifted in relation to political discourses of communism, capitalism, nationalism, justice and democracy.

In other words, we will see that problems stemming from the lack of intellectual or professional autonomy in Soviet society were not simply resolved by national independence. The hegemony and intellectual politicization of Soviet rule emerge as partial factors in the politics of Central Asian social science rather than as the defining factors. The relationship between organized knowledge and power has changed form as new forms of heteronomy creep into place. While there have certainly been ruptures between 'Soviet' and 'post-Soviet' science, there is also continuity in localized projects to both affiliate power with legitimate scientific knowledge and distance social scientific knowledge from 'illegitimate' power. Trends in knowledge production in Central Asia after 1991 are not reducible to either progressive reform or regressive underdevelopment. Rather, they are results of and responses to a new encounter between the logics of science, power and capital.

This goes some way towards demystifying the post-Soviet 'transition' in science and intellectual life (for a critique of this concept in political and economic contexts, see Liu 2003). It also departs from traditional narratives of the theme in which the crisis and collapse of Soviet communism were widely expected to usher in an unprecedented release of intellectual energy, and of interpretations that emphasize the 'modernization' or 'maturation' of academic disciplines. It has been widely assumed that the tension between knowledge and power, which is sometimes also articulated as a conflict between truth and politics or simply as the politicization of truth, would be resolved with the collapse of Soviet power. Indeed, 'the historical development of the social sciences is often seen in terms of a gradual liberation from traditional bonds which prevented them from realizing their full potential as producers of true, undistorted knowledge of society' (Wagner and Wittrock 1991: 3). Symbolically, if not always in practice, national independence was interpreted as a disruption in 'normal' Soviet science as defined within Marxist–Leninist philosophies of knowledge; a liberation of knowledge from power – a moment rather than a process of autonomization.

By the late 1980s, the organization of Soviet science was widely believed to be the primary source of its destruction. It was overdeveloped, hierarchical, centralized, politicized, intellectually isolated and relatively homogenous; in other words, antithetical to internationally dominant norms of scientific activity, and therefore a self-sufficient explanation for the stagnation and ideological role of organized knowledge in society (Ruble 1993: vii). From this point of view, it is logical to presume that adherence to norms of universalism, communism, disinterestedness and organized scepticism would correct for much of the distortion in Soviet scientific knowledge; that 'de-politicization', liberalization and democratization would correct the 'deviations' of Soviet science and bestow upon it greater legitimacy (see Merton 1996 for more on the ethos of science). This anticipated liberation represented more than a scientific revolution; for some, it confirmed the world-historical triumph of democracy and capitalism over totalitarianism and communism.

This narrative of intellectual progress still circulates widely in the region and underscores much academic reform that has been undertaken since the early 1990s. Scholars struggle to 'de-ideologicize' the subjective and objective conditions of their work. Survivals of the 'Soviet past' are identified, criticized and rooted out within the academic community (somewhat like 'national survivals' under the Bolshevik regime); innovations from 'the West' or 'civilized society' are sought, applied and revered. Marxist credentials are viewed with suspicion and professional training abroad (particularly in empirical method) has become a premiere source of cultural and academic capital. A general theory of transition-as-modernization summarizes this approach to knowledge and reveals its affinity to earlier evolutionary materialist theories of development: Central Asian societies are developing teleologically from communism to capitalism, totalitarianism to democracy, stagnation to development, backwardness to modernization and barbarism to civilization. Ideas – social science, organized thought – are not only destined to be transformed through this process but, as in all modernizing projects, are obliged to play a role in its realization as well.

This discourse also underpins debates about the nature and role of social science in Central Asia, particularly vivid surrounding sociology in Kyrgyzstan. On the one hand, the field has been confidently advertised as a 'vital element of democratic societies' in both academic and public venues (Isaev 1993; Isaev et al. 1996; Lokteva 1991). Sociologists have argued that the knowledge can be produced to serve 'the people' as opposed to a minority political and economic elite (Blum 1991; Isaev 1991a); that it is a 'path to development' (Isaev 1993, 1998a, 2003); that it can be a source of reliable and true information about society for people living through a period of collective unease and insecurity (Bekturganov 1997; Blum 1990; Isaev et al. 1994d); that it can and should be employed in the service of human freedom and not social control (though these are not necessarily mutually exclusive in this context) (Isaev 1998a; Isaev et al. 1994d); and that it is possible to establish a 'national sociology' which meets 'modern' and 'international' standards but that also accounts for the sociohistorical specificities of Kyrgyz history and culture (Isaev 1993b; Isaev et al. 1994b; Ryskulov 1998). Through this rhetoric, Kyrgyzstani sociologists have elaborated a mission to rescue legitimate truth, as it were, from the abuses of illegitimate power; to transform the practice of power-writing-truth-as-ideology into one of scientists-managing-truth-as-power. It is a modernization thesis and repressive hypothesis rolled into one; a project to, as one Kyrgyz journalist wrote, transform the production of illegitimate 'truth in strength' into a social politics legitimated by 'strength in truth' (Blum 1993). Although based on a study of Kyrgyzstani sociologists, this reflects a common experience across the social sciences throughout the region (Bronson et al. 1999; Salehi-Esfahani and Thornton 1998; Toshchenko 1998).

However, the actual status of social scientific knowledge and its appropriate relation to politics are contingent and controversial in practice. Many people – sociologists included – fear that its 'subjective' and 'unscientific' nature make it uniquely susceptible to ideological manipulation. At the same time, there is a widely held belief that fields such as sociology actually have tangible influence on social consciousness and action; that they have a 'direct impact on the formation of public opinion' (Bekturganov et al. 1994); that, in the words of one sociologist, 'the trajectory of planetary movement does not change depending on the progress of astronomers, but the level of sociological knowledge, as world experience shows, actually influences the direction and result of social transformation' (Isaev 2000).

This dual perception of risk and relevance has caused heated controversies about the need to demarcate 'positive' from 'negative' incarnations of social scientific knowledge.

It has also led to the rise of what Herbert Marcuse (1964) termed 'ideological empiricism', or the methodological exclusion of critical perspectives and political questions from social research. The purification of subjectivity and intentionality from social scientific inquiry is seen to be a historically proven answer to the power/knowledge problematic. A new breed of positivist and empiricist sociology has become central in attempts to reconcile these dual demands. Like its Marxist–Leninist predecessor, the new philosophy of sociology is defined as true, universal, objective and normatively neutral; it stands in contradistinction to 'ideology' and 'politics', which are seen to be false, particular, subjective and interested. Since independence, 'sociology' and 'Marxism–Leninism' have swapped roles in the power/knowledge debate: yesterday's truth becomes today's ideology, and vice versa. The post-Soviet revolution in social science – its 'liberation from class ideology' and Communist Party domination (Isaev 1998c) – may also be interpreted as a more moderate shift in substantive emphasis. In this quest for truth and legitimacy the politics of truth are deliberately bracketed, not least of all because the sociology of knowledge has its roots in the Marxian tradition of ideology critique that has, since the Soviet collapse, been declared 'unscientific'. The demise of Communist Party hegemony and Soviet rule over social science, as well as the subsequent decentralization of power and the predominance of neoliberal ideologies, are thus taken to mark the end of the power/knowledge problematic in Kyrgyzstan.

In fact, the embracing of universalistic and positivist epistemologies (and, as will be discussed, Orientalist and Eurocentric criteria of intellectual legitimacy) in a period which has elsewhere been defined as fragmented and particularized, postpositivist and global is a problematic of these very politics. It challenges illusions that our era is genuinely postmodern and raises questions about the status of knowledge communities on the edge of what might still be usefully defined as a global science system (Alatas 2003; McDaniel 2003). This raises further questions about the geopolitics of knowledge in post-Soviet space. Localized and deeply political debates about the nature and use of social scientific knowledge have dovetailed with new discourses and practices of 'development', particularly those which tie science and education to economic development, on the one hand, and those which embed cultural work in foreign policy on the other. Knowledge reform has become incorporated into international development programmes and framed by a grand narrative of progress and modernization, which posits an unproblematic transition from authoritarian state socialism to neoliberal capitalism. Development agencies and foreign governments have spent millions of dollars on projects to liberalize research and higher learning by disengaging them from the state and re-institutionalizing them in civil society or broader capitalist markets. In the wake of economic and political decentralization, particularly in the poorest regions, many schools, universities, research centres, arts organizations, publishing projects, professional training programmes and intellectual or materiel exchanges owe their existence at least in part to the sustenance of this development aid. The results, as illustrated by the case studies in this book, have been ambiguous and complex.

This does not minimize the advances that have been made towards expanding the critical and practical promise of scientific work in and about post-Soviet societies. International aid, sponsorship and support have virtually sustained social science in the region, providing resources, space and protection for types of research and teaching that would otherwise be unthinkable.[3] Social scientists have recently forged new paths in postmodern, feminist and neo-functionalist social theories; some are experimenting with Marxism again. The opening of national borders, archives and social relationships has enabled non-Soviet scholars to study some of the most elusive societies of the twentieth century; they in turn have produced findings that challenge the entire category of *homo Sovieticus*. Great swathes of historical experience have become visible for the first time through archival research and oral history, and the once-monolithic 'Soviet Union' has been replaced with a multitude of cultures and subjectivities, nuances and complexities, and examples of grassroots resistance to human massification.

However, these advances do not justify the uncritical and often ideological correlation between the Soviet collapse and scientific progress or intellectual production in the region, nor should their ostensibly progressive nature exclude them as possible subjects for critical analysis. The heteronomy of knowledge and the complicated relationship between knowledge and power is not merely a question of Soviet rule or independence, not simply one of 'de-ideologicization', 'de-politicization' or 'modernization', but rather rooted in a deeper ontological problematic inherent in the embedded nature of knowledge itself and the situated hermeneutical status of knowledge production in any society.

Understanding the politics of knowledge in former Soviet space, therefore, demands more nuance than can be achieved by measuring the pace of its 'transition' or its 'development' in relation to universalized standards of education, science and intellectual life – though this is often how the issue is framed 'on the ground' by indigenous academics themselves, who are understandably more concerned with the more ontologically pressing problems of job insecurity, poor working conditions and remuneration, lack of intellectual property rights, limited access to resources and alternative ideas, fraught with professional ethics and censorship. Some will be addressed, I hope fairly, in more detail throughout in the book. I argue, however, that the immediate problems facing social scientists in the region can and should be understood within broader historical and theoretical frameworks.

One major link that must be made in analyses of the politics of social scientific knowledge in Soviet and post-Soviet Central Asia is to colonialism and empire. This includes both Soviet colonialism and Western imperialism in Central Asia (Bruno 1998; Cavenaugh 2001; Clem 1992), and requires that we recruit theoretical concepts and empirical comparisons from earlier literatures on postcolonial knowledges (e.g. Ake 1982; Eisemon 1982; Fernandes 1967; Joshi 1995; Wagner and Wittrock 1990). We have witnessed, alongside the emergence of new academic freedoms and professional possibilities, the imposition of new forms of intellectual and politico-economic domination in social science across the post-Soviet world. As concern wanes about the impact of Soviet imperialism on scientific knowledge, we must refocus our gaze on two other sources of heteronomy that have gained strength in its wake: the global 'market' and the nationalizing state.

The Soviet Union was a colonial society based on a deliberately maintained differentiation between centre and periphery, in the organization of science as much as in other social institutions.[4] However, the development of social science in Central Asia, along with the establishment of educational and scientific institutions in the region in the early twentieth century, has until recently been interpreted as the first stage of modernization and scientific enlightenment, and contrasted favourably to the scientific 'backwardness' of the Central Asian people prior to their incorporation into the Soviet empire. Furthermore, even on the periphery, 'Soviet science perceived itself as being the center' of world science (Nesvetailov 1995: 66). While it is important to acknowledge the significance of this identity for many scholars working in remote regions of the former empire, it is also important to recognize that the assimilation of the history of republican social science into the grand narrative of Soviet 'civilization', the use of Russian standards and categories as yardsticks of social development and national pride in Kyrgyzstan, historical amnesia about the repression of alternative historical narratives and post-Soviet counter-reactions to each of these tendencies may all be seen as consequences of the colonial logic of social science in Central Asia. The history of social science in the region must therefore be analysed not in a narrow national context, but within the broader framework of the Russian colonization of Central Asia, its continuation in the politics of Sovietization and the subsequent relations of 'development'.

The story told in this book therefore raises wider questions about the relationship between social scientific knowledge, the state and the market in the postsocialist world. It demands that we broaden our imagination about the meaning of this concept and see the postsocialist not only as a place or an era but also as a historical milieu, as a 'complex political, social and intellectual transformation brought about by the collapse of the "socialist" alternative to capitalism' (Outhwaite and Ray 2005; Peet and Watts 1993; Robinson 1996). This condition is not only characterized most generally by the collapse of geopolitical bipolarity, rise of a new neoliberal hegemony and – though a contested point in this book – the collapse of modernist grand narratives about social development but also by the institutionalization of a sense of 'choicelessness' about or determinedness of the trajectory of these and other changes (Sakwa 1999: 15). These changes impact upon the organization and imagination of social science in the world's newest independent nations, and upon how social scientists on the periphery of the global science system conduct their everyday professional practices.

9

This interpretation opens up a whole range of possible theoretical perspectives on Soviet and post-Soviet knowledge production and allows for serious research into the politics of knowledge in the region. Foucauldians will instantly recognize the repressive hypothesis in the 'from truth in strength to strength in truth' formulation; its critique is indeed a major concern of this book. Those familiar with the sociology of scientific knowledge will also quickly identify asymmetrical 'empiricist' and 'contingent' repertoires at work in defining what knowledge is and is for; rhetoric is part of science-as-usual rather than an exclusively post-Soviet phenomenon (Harris 1997; Locke 2001; Potter 1996; Taylor 1996). Studies of boundary-work among Central Academics reveal new insights into just how contingent notions of 'science' and 'truth' actually are. And while earlier studies of colonial and neocolonial knowledge production have been somewhat eclipsed by more recent attention to subalternity, hybridity and knowledge networks, the earlier and more 'modernist' categories remain useful for understanding why the fate of organized knowledge has historically been so politically and emotionally charged in the societies that once comprised the Soviet Union.

Classical questions in the sociology of knowledge are also invoked here. Are the meanings of 'science' and 'truth' universal or culturally and historically contingent? Is functioning democracy a precondition for scientific institutions, or is the political economy of truth genuinely contingent? Must 'power' and 'knowledge' be seen as mutually exclusive phenomena, as in the positivist tradition, or mutually constitutive, as in the post-structuralist? Is the quest for autonomous truth still a valid goal for social scientists to pursue? Can it be achieved without disregarding the contribution of critical sociology, the dialectics of subjectivity and objectivity in research and the workings of cultural power in society? And perhaps most importantly for many Central Asian social scientists, can social science be simultaneously 'political', 'moral' and 'scientific'? To what extent can it reveal the 'objective truth' about 'social reality', and can such knowledge, if it exists, contribute to the improvement of social life? Is there still a place in social science for truth?

Many social scientists today, including those working in critical and radical traditions that prioritize reflexivity, fail to appreciate just how sentient these questions are for our everyday intellectual practice. It is not until our *doxa* is disrupted, not until we doubt our own best answers – or are told that they are categorically flawed – that we begin to ask the most fundamental and difficult questions about what we know, how and why. At one of the outermost edges of the former Soviet Union, however, these foundational questions about the meaning of social science are raised, debated and negotiated in the daily exercises of postcolonial professionalization and academic practice.

Organization of the book

The remainder of the book is devoted to theorizing this experience through the critical theory of social scientific knowledge. Chapter 1 disarticulates narratives

10

of post-Soviet 'transition' from the normative political and epistemological assumptions in which they are embedded and focuses on how concepts of autonomy and heteronomy may be used to understand the political economy of knowledge in both communist and capitalist societies. Chapter 2 continues with a social history of social science in Central Asia prior to independence. It situates Central Asian science within broader histories of colonialism (Russian, Soviet and neoliberal), focusing on concepts of intellectual colonialism, academic dependency and Orientalism/Occidentalism. In Chapter 3, the discussion of colonial science is grounded in an exploration of the emergence of sociology in Central Asia. It departs from official Soviet histories of the topic to present a view from the periphery and explains how they defined acceptable relationships between social scientific knowledge and political practice within the Soviet science system. Chapter 4 narrates the further articulation of this politico-scientific ethos and its affiliation with 'national' politics during perestroika. Set in the context of movements for republican autonomy within the Soviet Union, this chapter explores how Kyrgyzstani sociologists used this new political atmosphere to further their own professional project, advocating the importance of intellectual independence within the context of a more critical Marxist ideology. The impact of independence on social scientific work is discussed in Chapter 5. First presenting the initial hopes and expectations raised both within and beyond the Soviet Union for the expansion of intellectual freedoms and scientific progress, the chapter then analyses how academics responded to these sudden changes by shifting the boundaries of science to exclude previously valued Marxist concepts and include previously marginalized capitalist agendas, while maintaining a uniquely Soviet ethics which posited a necessary link between scientific truth and political praxis. Chapter 6 illustrates this process through case studies of social science reform in the post-Soviet period, revealing how macro-level processes of marketization and political transition impact upon, and have been mediated through, the professional practices of academics in Kyrgyzstan. The de-socialization of knowledge is therefore shown not to have had uniform effects, but creates different constellations on the uneven playing field of scientific institutions in the region. This is continued in Chapter 7 which highlights how the boundaries of social science in post-Soviet society have historically been made into issues of public science. Two major debates about scientific truth which were published in the Russian- and Kyrgyz-language media are used to examine how a new ethos of science is being carved out in this space, and to explain why this is important as a socio-political, not merely academic, problem. Finally, the conclusion recapitulates the key argument of the book: that social science in Central Asia has historically been shaped not only by external structures of domination and liberation but also by academics' actions to negotiate the boundary between truth and politics in order to create a discipline that is both scientifically true (according to the logic of science) and politically normative (meaningful within the logic of power). It concludes that the intellectual emancipation promised by political and market reforms in science and education is being eroded through the colonization of the academy by the demands of the nationalizing state and by market mechanisms such as inequality, commercialization, competition and supply and demand within the scientific community itself. The concluding chapter ends by tying post-Soviet social science back into broader questions in the sociology of knowledge and trends in the organization and role of science within the global system.

1

KNOWLEDGE AND POWER IN POST-SOVIET SPACE

In our state, sociology must be a science and not an ideology.
(Bakir Uluu 1997)

The study of the truth must itself be true.
(Isaev in Sydykova 1998)

What are the social conditions which must be fulfilled in order for a social play of forces to be set up in which the true idea is endowed with a strength because those who have a share in it have an interest in truth, instead of having, as in other games, the truth which suits their interests?

The fundamental question of the sociology of science assumes a particularly paradoxical form in the case of the social sciences: what are the social conditions of development of a science freed from social constraints and demands, given that, in this case, progress in the direction of scientific rationality does not mean progress in the direction of political neutrality?
(Bourdieu 1975: 31, 36)

Debates about the politicization of social science and interest in the political nature of knowledge itself are not particular to post-Soviet societies, nor were they absent in the Soviet history of science. Intellectual projects to delineate epistemological boundaries between 'knowledge' and 'power' and to establish legitimacy for competing visions of social science have shaped the history of the field; the *Streit um die Wissensoziologie* ['sociology of knowledge dispute'] in early twentieth-century Germany (Meja and Stehr 1990), the mechanism–dialectics debate in Soviet philosophy during the same period (Sheehan 1993), and the more recent 'science wars' in the US and the UK (Segerstråle 2000) are only three examples. The current politics of social science in Central Asia may be understood within the context of such recurring debates in the sociology of knowledge, which pivot on questions about the ontological nature of knowledge, the philosophical possibility of knowing the 'truth' about society and the viability of professional attempts to create institutional conditions which might enable such knowledge to

13

emerge. Embedded within each of these are more focused concerns regarding the relationship between organized science and other social institutions within a given politico-economic system, and between different fields within knowledge communities or institutions.

The dissolution of the Soviet empire and subsequent processes of decolonization in former Soviet space have created a new social context for the re-emergence of these foundational debates in the sociology of knowledge. The relationship between knowledge and power in the post-Soviet world is of interest not only because it shapes the meaning of scientific knowledge in cultural institutions but also because it offers comparative insight into the sociology and politics of organized knowledge in general. Current debates about the politics of social science in Central Asia must therefore be situated within three broader contexts: the sociology of knowledge and science, recent trends in organized knowledge and 'globalization', and postcolonial comparisons in the sociology of knowledge.

Philosophy of knowledge: a view from Central Asia

Discussions about social science are often not framed in such terms by Central Asian academics. They have reflected on this theme since late socialism, often as a debate of knowledge *versus* power (Blum 1990; Fanisov 1990; Isaev 1991, 1991a) or knowledge *for* power and as a *technique of* power (Abazov 1989; Sorokina 1989; Tishin 1980). However, it is seldom considered that the fields of power and knowledge may be mutually constitutive, or that the nature of disciplinary institutionalization has been shaped by attempts to distinguish them or otherwise reconcile the logic of science with that of power.

In Kyrgyzstan, the emergence and formation of the sociology in particular has been framed primarily as a problem of disciplinary maturation and institutionalization: a functional process measured quantitatively and qualitatively at the level of formal institutions. There is much discussion about when and where research laboratories and centres were established in educational, industrial and political institutions; when decrees relating to science and academic work were passed by the state or the Communist Party; when a new theme was published or taught; when a conference was held. Progressive stages of institutionalization are measured by counting the number of departments, institutes, students, associations and publications that exist (Isaev 1993; Ismailova 1995; Zarlikbekov 1998), enumerating indicators of credibility such as how much legitimacy sociology is afforded by political leaders, development organizations and scholarly associations (Blum 1993; Isaev *et al.* 1993a), and drawing comparisons of indigenous or 'national' sociology with the discipline in its more 'mature stages of development' in the US, France, Germany and Russia (Blum 1993; Isaev 1993).

The following excerpt from a newspaper article entitled 'Problems with the institutionalization of our sociology' (Isaev 2000) represents the prevailing view on what factors facilitate the institutionalization of social science:

> The establishment of sociology as an independent sphere of scientific knowledge, an academic subject and a profession is difficult for us. This is seen in the lack of a special scientific-sociological knowledge and way of thinking, trained cadres and traditions, and the long-term dominance of the ideologicization and politicisation of quasi-social scientists.... As shown by the experience of other countries, the institutionalisation of sociological knowledge depends on the appearance of specialist – professionals, the achievement of a mature status, the formation of a particular infrastructure, a calling to support the reproduction and translation of knowledge, investment in scientific associations, and etc.

Missing, however, are explanations of the social factors that make these particular phenomena possible in the first place, such as structural relations between science and socio-economic institutions, the impact of international relations, the transfer of financial dependence from the state to international organizations and *zakazchiki* [commercial clients], and more epistemological issues such as the perceived relationship between truth and power, the role of social science in society and public understandings of 'the scientific'. It is also grounded in a number of unexamined assumptions about scientific knowledge and its relationship to power: that national independence is the most logical point of departure for understanding postcoloniality; that academic disciplines must be defined and measured according to a pre-existing set of criteria. What constitutes a 'sociological way of thinking' and how does it become shared or guarded among a group of practitioners? Are the conditions which make possible the emergence of sociology as an academic discipline the same as those that make possible its development as a science or applied profession? Who or what determines when a field has reached maturity? These and other questions remain unanswered.

Narratives of linear and progressive liberalization often obscure the complex processes by which different conceptions of truth are legitimized in society. In Central Asia, the adoption of this approach is made even more problematic by the fact that it is a stylized representation comprising a number of socially specific cases in the history of social science (e.g. German, British, American and French sociologies), which are conflated and redefined as universal within an Occidentalist discourse of 'the West' and sanctified by ideologies of modernization and development.[1] This ideal-type model, however, is meaningful for many scholars in the region who believe the Soviet legacy *per se* to be the enemy of intellectual and professional progress and 'scientific sociology' to be its salvation.

The way forward appears unproblematic: social scientific knowledge can and must be divorced from power and made objective, as it is assumed to be in the 'civilized countries of the world' (Sydykova 1998). It seems an obvious solution for scholars living in societies that define themselves, if often tongue-in-cheek, as 'democratizing'. Once considered the handmaiden of Soviet power, social science has been redefined as a 'scientific' defence against the abuse of knowledge by opportunistic academics and elites, being democratic in its ostensibly neutral methodologies and 'equidistant from all power structures' (Isaev 1998c; Isaev *et al.* 1994b). In this discourse, the presumed power of social scientific knowledge shifts from being owned and managed by the *verkhnye* ['upper' or ruling class] to being directly accessible to the *nizhnye* ['lower' or people] by providing them with impartial 'information' that they can use to make personal and political decisions, and making political leaders accountable for their actions.

The current predominance of positivist approaches to sociological research in Kyrgyzstan is thus partly a response to social demands for certainty. It is also rooted in an epistemological realism common to both Marxist and positivist theories of knowledge. This position posits that there is a stable, ontologically existent and empirically apprehensible distinction between 'ideology' (distorted or mystified knowledges and false consciousness) and true, realistic and objective 'knowledge', also synonymous with truth (Gieryn 1983: 783; Kitching 1994; Lincoln and Guba 2003). 'Genuine' science is seen to belong to the latter category, and may be distinguished from 'non-science' through the correct delineation of boundaries between knowledge and illegitimate power, which distorts truth for purposes of exploitation or domination, and a mutually constitutive relationship between knowledge and legitimate power, which both ensures the production or discovery of objective truth and enables it to be used for social good. Hence, by dispensing with all ideologies, social scientists can obtain true and undistorted images of social reality, which can in turn be applied by 'experts' to make political decisions more scientific and therefore make social science more politically effective (for more on the 'positivist persuasion' see Alexander and Colomy 1992 and Kuklick 1980). The development of sociology is then carefully measured against a universalized ideal of 'mature' social science, in the likeness of Comte's ([1853] 1975) social physics or Durkheim's (1938) sociological method, represented during the Soviet period by sociological practice in Moscow and Leningrad, and today by sociology in 'the West'. Here, the definition of 'mature science' is synonymous with 'autonomous science', the evolution of a methodology for knowing social reality that is able to transcend the phenomenological subjectivities of politics and experience in order to contribute to the governance of both.

There seems to be little reason or room for questioning this position. While the ideologies of Marxist science have been dislodged from social science in Central Asia, they are being replaced with positivist and functionalist philosophies of science that are in certain ways epistemologically compatible. The roots of faith in the possibility of scientific sociology extend equally deep into the realism of both Marxist and non-Marxist philosophies. For example, while Marxist–Leninist and functionalist philosophers have levied totalizing critiques against one another, there has also been considerable attraction to and work around positivist structural-functionalism in the region since the mid-twentieth century (Vucinich 1974). In both cases, discursive dualities have been used to legitimize particular definitions of social science: while Soviet Marxist sociology is now defined as 'speculative' and 'ideological' in relation to non-Marxist positivist and functionalist methods, it was precisely these categories which were previously used to distance Marxist social research from both 'bourgeois' empirical sociology and abstract functionalist philosophy (Oruzbaeva 1980; Sheehan 1993; Vucinich 1974).[2]

This epistemological common-ground, which often goes unrecognized, has facilitated the interweaving of Marxist and positivist philosophies of knowledge in new articulations of sociology in post-Soviet space. Positivist theories of social fact and empirical research techniques have become attractive not only because they are believed to be immune from theoretical bias but also because they are said to reveal underlying social 'laws' and 'regularities'. The structural theory of a natural-law-governed social reality has been combined with an empiricist theory of the primacy of observed reality by detached observers. In postpositivist circles, this may easily be interpreted as evidence of a resurgence of 'naïve' or even misguided positivism (e.g. Fisher 1990: 5). Far from being naïve, however, positivist vocabularies are also used by Central Asian sociologists in subversive ways to cloak potentially 'dangerous' social and political criticism in a mantle of scientific neutrality which makes them less susceptible to reprisals from the subjects of their research, which are often national power elites – after all, you 'can't blame a mirror' for what it simply 'reflects' (Blum 1991; Fanisov 1990). Critiques portrayed as representations of objective fact leave little justifiable space for accusing scientists themselves of constructing falsehoods.

Given the deep epistemological and political moorings of the positivist position on knowledge and its subsidiary function as a rhetorical device for social criticism, critique of this first principle of the realism or naturalism of scientific knowledge is often interpreted as a challenge to the legitimate authority of social science or even an assault on truth itself. This became abundantly clear to me during an interview, with a well-known sociologist in Kyrgyzstan, conducted in 2003. I presented him with two newspaper articles about sociology written in recent years. One was entitled, 'Who benefits from populism in sociology?' and the other, 'A sociology of lies, or the lies of the sociologist?' Each was part of a longer-running public debate about the proper relationship between sociology and politics, the importance of institutionalizing the scientific method in social science and the collective professional response to data manipulation within the academic community. The professor, a long-time campaigner for sociology in both the Soviet and post-Soviet periods, disagreed that the articles concerned the 'politics' of sociology. He argued that these contributions to the debate were anti-political; that it was an attempt to demonstrate how other people had politicized sociology and to appeal for science to be 'pure' and free from political influence. In other words, he did what scientists tend to do when accounting for discrepancies between their truth-claims and others' – couch his own position in an 'empiricist' rhetoric of scientificity and objectivity and represent competing positions in a more 'contingent' discourse which emphasized subjectivity, bias and incompetence (see Mulkay and Gilbert 1982). This 'asymmetrical' association of the political with illegitimate power and particular ideologies, and of science with universal truth means that the actual politics of sociological knowledge themselves go largely unexamined.

Such rhetorical devices of course have more than symbolic significance. The embracing of realism, positivism, empiricism and modernist ideologies of science must also be understood as a considered and situated response to contemporary uncertainties about the role of knowledge in the society: what constitutes legitimate social knowledge and who qualifies as a legitimate knower, if and how this knowledge can be applied to political decision-making and social development, whether social information is relevant to human development, and what the ultimate goals of a science of society – indeed, of that society – should be. In order to establish legitimacy for their work, Central Asian social scientists are pressed to prove that it is *not* susceptible to political perversion, but that it *is* politically relevant to public concerns. This is the paradox that Bourdieu speaks of in the quote earlier.

The intersection of sociology, knowledge and power means that the development of sociology in Central Asian societies must not be taken for granted as 'inherently progressive and truth-producing', as it is often presented in reformist discourse (Beliaev and Butorin 1982; Popkewitz 1991). Social science reform must be analytically defined as the practice of institutionalizing new relations between organized knowledge and socio-political power in both intellectual and institutional forms (Ake 1982; Bujra 1994; Eades and Schwaller 1991; Gosovic 2000). The two-pronged nature of the project – to affiliate power with knowledge (in the construction of communist society during the Soviet regime and in the realization of capitalist reforms after independence) and to separate social scientific knowledge from 'illegitimate' power – makes it difficult to untangle the mutually constitutive relations between sociology and power at various stages of the discipline's historical development. The study of knowledge and power in Central Asian social science is necessary not in small part because it is the relationship most often neglected in post-Soviet sociology itself.

Sociology of knowledge

The sociology of knowledge offers an alternative framework for understanding knowledge production in Central Asia, one that allows for consideration of the relationship between knowledge and power in social science. It takes foundational assumptions about truth and objectivity as its primary problematic, seeking to explain how and why 'certain forms of truth come to prevail, and be challenged, at different historical moments' (Popkewitz 1991: 43); clarifies the 'conditions under which problems and disciplines come into being and pass away' in any society, particularly those experiencing great change (Mannheim 1936: 97); and enquires into the 'role of knowledge and ideas in the maintenance or change of the social order' (Wirth 1936: xxx). In the post-Soviet context, the primary questions about sociology from the perspective of the sociology of knowledge are therefore not why the discipline has failed to mature, or what measures are needed to make it 'world class' (Isaev 2000), but rather how these conceptions of maturity and standard gained cultural currency and what accounts for the ascendance of a new type of orthodox positivism in what is often presumed to be a postpositivist period in the history of knowledge (Feyerabend 1981, 1993; Kuhn 1970; Lakatos and Musgrave 1970; Vasquez 1995).

The sociology of knowledge is a broad field, ranging from Karl Marx and Friedrich Engels' (1991) theory of ideology and Karl Mannheim's (1936) classical study of the 'relationality' of individually (i.e. class) positioned knowledges, to the 'new sociology of knowledge' that is more gestalt-oriented in its focus on how the whole epistemic apparatuses are created (Swidler and Arditi 1994: 306). On and beyond this spectrum lie works from theorists as diverse as Hannah Arendt, Pierre Bourdieu, Emile Durkheim, Michel Foucault, Donna Haroway, Jürgen Habermas, Max Horkheimer, Thomas Kuhn, Imre Lakatos, Georges Lukásc, Robert Merton, Edward Said, Max Scheler and Dorothy Smith – a small sampling of philosophers and social scientists who have, sometimes in contradictory ways, concerned themselves with the social origins and consequences of knowledge. What unites them is the basic proposition that social thought 'has an existential basis insofar as it is not immanently determined and insofar as one or another of its aspects can be derived from extra-cognitive factors' (Merton 1996 [1945]: 209). The sociology of knowledge is based on a constructivist epistemology and asserts that, to varying degrees, knowledge-claims are social and political productions, cultural phenomena – constructed, produced and negotiated, not things-in-themselves to be objectively discovered – and that the ideational world is not independent from the material conditions within which it is created (McCarthy 1996). This holds equally true for lay knowledge and 'expert' opinion, academic science and common sense: all ways of knowing are subjective and 'political' because they are situated in human experience and social relationships.

The sociology of knowledge thus calls for the absolute democratization of ideology critique, whereby all positions and truth-claims may become subjects for critical analysis (Mills 1963a: 457). Sociologists of scientific knowledge refer to this as 'symmetrical' analysis, which seeks naturalistic causes of legitimate truth-claims as well as those of 'error' or falsification (Bloor 1991). This perspective demands that we rethink the definitions of both 'ideology' and 'truth' and make a heuristic shift from what Mannheim referred to as a 'particular' to a 'total' conception of ideology in order to move beyond the critique of 'other' truth-claims and understand the underlying logic of entire debates. Particular ideology is defined as the 'more or less conscious disguises of the real nature of a situation, the true recognition of which would not be in accord with [one's] interests'. Total ideology, on the other hand, is what Mannheim refers to as 'the ideology of an age or of a concrete historico-social group, e.g. of a class... the characteristics and composition of the total structure of the mind of this epoch or of this group' (Mannheim 1936: 49). While the former allows for the possibility of existentially autonomous knowledge, the latter asserts that even knowledge that has not been politically 'corrupted' may be embedded in broader power relations. There has been much critique of these categories, particularly for their ideal-type generality (Bailey 1996; Horkheimer (1990 [1930]); Meja and Stehr 1990). However, they remain useful for establishing a basic understanding how organized knowledge itself may become an object of sociological inquiry.

The distinction between particular and total ideology, or between ideology critique and the sociology of knowledge, is particularly important in Central Asia, where a particular ideology of the superiority of social scientific knowledge, as well as an emerging professional ethos of 'neutral' scientific method, has developed in response to the politicization of organized knowledge under the Soviet regime. But within the sociology of knowledge, even the truth-claim of de-ideologicizing truth must be critically examined.

For many years in Western European sociology, scientific knowledge was considered to be largely immune from the sociology of knowledge, the debate continues in 'science studies' amidst a newer set of theories about the social construction of science itself (Bloor 1991, 1999; Latour 1999a). When taken to its logical conclusion, however, Mannheim's theory of ideology suggests that we must subject even the foundations of social scientific knowledge to sociological analysis. This is not implied in his original work; in fact, he believed that a certain type of transcendent knowledge might be produced by economically and socially dislocated intellectuals who, being unattached to any specific class, would have a multi-perspectival view of all truth-claims in a given society (Mannheim 1936). Before and after this seminal work was published, however, other theorists were less confident about the 'special' status of intellectuals and scientists. In the words of Robert Merton (1996 [1945]: 207), 'the sociology of knowledge came into being with the single hypothesis that even truths were to be held socially accountable, were to be related to the historical society in which they emerged'. For

> as long as one does not call his own position into question but regards it as absolute, while interpreting his opponents' ideas as a mere function of the social positions they occupy, the decisive step forward has not yet been taken.... In contrast to this special formulation [of particular ideology], the general form of the total conception of ideology is being used by the analyst when he has the courage to subject not just the adversary's point of view but all points of view, including his own, to the ideological analysis.
> (Mannheim 1936: 68)[3]

20

This reflexivity stimulates the development of a more specialized branch of sociological inquiry: the analysis and critique of scientific truth itself. The main premise is elementary, though highly contested even within the field: that science is also a socio-cultural phenomenon and neither its organization nor products can be fully understood outside an analysis of its surrounding socio-political and cultural contexts and of the internal organization of intellectual activity itself (Blume 1974; Bourdieu 1975: 19; Reynolds and Reynolds 1970; Smith 1988). The sociology of scientific knowledge takes science itself as its main object of sociological inquiry, which, like any other body of knowledge, 'cannot be adequately understood as long as [its] social origins are obscured' (Mannheim 1936: 2).

According to theorists such as Bourdieu (1975, 1988, 1991), Kuhn (1970) and Smith (1988) these 'origins' often include power relations that influence what becomes accepted as true and legitimate knowledge, both within scientific communities and in the public sphere. These are manifested not only in the overt attempts of political forces to exert control over knowledge production (or vice versa) or of social scientists to assert their professional autonomy, but may also be expressed in contests for position, resources and prestige within cultural institutions, as well as in the intellectual debates held at conferences, in journals, and in informal settings. Although there is no central problematic in the sociology of scientific knowledge, schools of thought can be broadly distinguished by how they classify these 'internal' and 'external' relationships (Cozzens and Gieryn 1990; Ram 1991; Shlapentokh 1987). A theoretically synthetic approach which explores the nuanced interrelations between the internal–external and social–intellectual dimensions of the field is therefore well suited to analysing the co-constitutive relationship between social scientific knowledge and social structures and forces. This enables examination of the socio-historical contingency of knowledge as opposed to positing its simple determination by or autonomy from internal or external forces (Cozzens and Gieryn 1990; Wagner and Wittrock 1991). In Central Asia, where intellectual work is deeply embedded in its socio-political contexts, such an approach is imperative.

Insights from post-structuralist sociologies of knowledge and power

The works of Michel Foucault and Pierre Bourdieu are particularly important in this research because of the insight each offers into the politics of knowledge production, specifically within the social sciences. Foucault's inquiries into the historical construction of psychiatry, deviance, criminality and the human sciences more generally exposed an intimate relationship between 'expert' knowledge, particularly that claiming the privileged status of 'modern science' and the exercise of repressive power in society (Foucault 1967, 1973, 1989, 2001). His work enables us to imagine how people come to accept as 'natural' truth-claims that are contingent and political, and why we often bestow legitimacy upon scientific classifications of experience which may under other circumstances be otherwise contrived.

Bourdieu's work in the sociology of social science and institutions of knowledge production (universities, the arts, literature) begins from a similar assumption, that 'the objective truth of the product – even in the case of that very particular product, scientific truth – lies in a particular type of social conditions of production, or, more precisely, in a determinate state of the structure and functioning of the scientific field' (Bourdieu 1975: 19). Whereas Foucault's work focused on looking at how authoritative truths are institutionalized by nebulous discursive power relations within society, Bourdieu's is more a structural theory of truth which looks concretely at how institutional structures and power relations – particularly those based on class and professional position – within knowledge-producing institutions shape socially sanctioned definitions of truth (Bourdieu 1975, 1988, 1993). For him, 'the "pure" universe of even the "purest" science is a social field like any other, with its distribution of power and its monopolies, struggles and strategies, interests and profits, but it is a field in which all these invariants take on specific forms' (Bourdieu 1975: 19). While maintaining that knowledge is produced through human agency, Bourdieu asserts that intellectual positions and professional strategies are 'structured' as well as 'structuring'. Distinctions and hierarchies in the academic world serve not only to legitimate certain truth-claims while stigmatizing or excluding others, but also represent power relations, which determine who is permitted to participate in and evaluate this cultural work. For Bourdieu, the quest for scientific truth in the academy is therefore inherently political even when not deliberately politicized:

> every scientific 'choice' . . . is in one respect . . . a political investment strategy, directed, objectively at least, toward maximisation of strictly scientific profit, i.e. of potential recognition by the agent's competitor–peers.
>
> (1975: 23)

Following Foucault and Bourdieu, it may be argued that the production and legitimation of social scientific knowledge are political, both in the way that contests for material and symbolic resources (including scientific authority) affect knowledge construction, and how that knowledge is politicized through its application. In these perspectives, power is a precondition to the production of truth-claims in scientific knowledge, not an obstacle to it (see Fuchs 1986). Confronting the issue of power/knowledge directly, Bourdieu argues that 'the idea of a neutral science is a fiction, an interested fiction which enables its authors to present a version of the dominant representation of the social world, neutralized and euphemized into a particularly misrecognizable and symbolically, therefore, particularly effective form, and to call it scientific' (Bourdieu 1975: 36). This, obviously, raises serious questions about any projects to separate out questions of scientific knowledge from those of power: if the theory is correct, these practices are in doubt; if the practices are effective, the theory must be re-examined.

22

Extending the sociology of knowledge to Central Asia

Social theories of the existential conditioning of knowledge have a number of important implications for the study of social science in Central Asia. At the most immediate level, because they compel us to examine the relational and contingent dimensions of all truth-claims about society, including those that claim to be scientific or objectively value-free, we can theorize projects to autonomize social science from what can only be described as a heteronymous state of existence. The claim to value-freedom itself becomes a site of critical sociological enquiry. Furthermore, the post-Soviet condition is particularly suited for studies in the sociology of knowledge, and vice versa. The sociology of knowledge has historically been employed to make sense of contradictory truth-claims in periods where 'disagreement is more conspicuous than agreement', to prevent one falling prey to either dogmatic absolutism or epistemological relativism, and in order to analyse the deep structural meanings of political rhetoric and ideology when these are naturalized as 'truth' in the public mind (Mannheim 1936: 5). It is particularly useful in circumstances where 'norms and truths which were once believed to be absolute, universal, and eternal, or which were accepted with blissful unawareness of their implications, are being questioned' (Blume 1974: 2–25; Wirth 1991). This is an appropriate description of contemporary perceptions of post-Soviet Central Asian society, which are pervaded by images of chaos and crisis. In fact, sociologists often portray society itself as a mystery, some dangerous *terra incognita*; the reading public is periodically reminded in ominous tones that 'we do not know the society in which we live' (Blum 1993; Isaev 1991a, 2000).

Such statements are used to increase public demand for social research in the region, but they are not merely rhetorical. The right to 'know one's society' here refers to expectations about modern forms of collective social knowledge which are routinely produced by social institutions such as schools, media, science and the academy, and which were routinely suppressed under the Soviet regime. While apocalyptic warnings about the dangers of ignorance categorically fail to acknowledge any of the non-institutionalized and primary ways that people 'know their society' – experience, common sense, relationship, sensation, reflection and mediation – it is a fact that organized knowledge in the form of social theory and research was censored and proscribed in Soviet societies. Hence, not only the present 'transition' but also past and future are experienced in terms of intellectual crisis.

This is part of a broader social crisis, which is represented as a negative and undesirable phenomenon, something imposed by the Soviet collapse and which worsened after independence. While early sociological articles in Kyrgyzstan chronicle the intellectual and political anxieties of perestroika (e.g. Isaev 1991, 1991a; Zhivogliadov 1990), those published after independence tend to be more critical of the severe poverty, heightened corruption, ideological anomie and everyday violence that followed national independence (e.g. Bekturganov 1997; Isaev 1993b,c, 2003) as well as the loss of public trust in knowledge-producing institutions such as the media and academy. The discrepancy between 'reality' and 'ideology' or lived experience and official rhetoric is a key problematic in sociological research conducted during this period (Yurchak 1997, 2003). The sociology of knowledge is equipped precisely to deconstruct the politics of truth in such conditions, where, as was written of Mannheim's Germany, we find both 'stunned disillusionment' and 'the rapture of a newly-won contact with true reality', or at least of its potentiality (Kecskemeti 1952: 2).

In addition to the emergence of competing truth-claims, however, there is also in Central Asia an effort to debunk existing norms and belief systems (e.g. 'the socialist way of life') and replace them with others (such as concepts of 'the nation', 'democracy' and 'the free market') that are not clearly or unanimously defined. This is visible in discourses on 'the transition' in Kyrgyzstan, which, in opposition to 'the crisis' is defined as a clear and determinate progression from a 'totalitarian' and 'communist' society to a 'democratic' and 'capitalist' one (Isaev 1995; Isaev and Madaliev 1998; Isaev *et al.* 1994e). The transition is the optimistic discursive counterpart to the crisis; an ideological promise of inevitable things to come, the vision of the foretold future. Because the particular theory of development underlying images of 'transition' is evolutionary and teleological, neoliberal models of democratization and capitalist transition are offered up as the most logical solutions to the crises of independence (see Burawoy 1992; Liu 2003). Triumphalism and confidence about them dispels feelings of pessimism and disillusionment, which in practice have translated into social apathy and political indifference; in fact, they allow arguments that such reactions are 'normal' parts of depoliticization and de-ideologicization (Isaev and Madaliev 1998).

The emerging discourse of transition therefore compels us to seek explanations of intellectual consensus-building and creation as well as of fragmentation and disintegration. Here too the sociology of knowledge can help us to understand the processes of knowledge construction and truth validation, for it asks specifically why certain forms of truth arise and are contested at particular historical junctures. This is vitally important in Central Asia, where the abandonment of one set of truths and assumptions (originating in Marxist philosophies of science and society) and acceptance of another (those embodied in discourses of positivist and empirical social science) have intersected with each other to create powerful discourses of scientific sociology for the post-Soviet age.

Boundary-work and the construction of knowledge

Approaches to the sociology of knowledge which bring together questions of academic knowledge, political hegemony and professional practice are particularly useful for analysing how the meaning of knowledge has changed as social science shed its Soviet affiliations and was realigned with the rhetoric and realities of national sovereignty. The analysis of boundary-work is one such approach. Boundary-work is used in three circumstances: when members of a discipline want to extend their authority into domains claimed by other professions or disciplines, monopolize professional authority and resources and see other disciplines or practices as 'competitive' or rivals, or protect their autonomy from encroachment from external sources (Gieryn 1983; Gieryn *et al.* 1985). This definition of boundary-work as expansion/monopolization/autonomization is grounded in particular assumptions about the political economy of science itself, however, and adheres to a 'market model of professionalization' in which academics use rhetorical strategies to create and monopolize new 'markets' for professional services in conditions where both material and social capital are scarce and in demand (Gieryn *et al.* 1985; also Bourdieu 1975, 1993). The popularization and institutionalization of a legitimate knowledge field are thus interpreted not as signs of the victory of truth over ignorance or ideology, but rather as indicators of a successful occupational monopoly within a competitive market. They reflect not possession of ultimate truth, but on how convincingly practitioners can argue that they offer an exclusive, relevant and legitimate view of the world. The ability to determine and monopolize the field is in turn influenced by the social positions of the actors themselves; the logic of scientific discovery is pre-politicized by its very position within and relation to larger structures of social divisions and power in society.

Applying boundary-work to sociology in Kyrgyzstan

The case of Kyrgyzstani sociology is a particularly useful illustration of boundary-work because the three main types outlined by Gieryn (1983) have often been conducted concurrently. In the 1960s, for example, academics promoting the establishment of sociology in Kirgizia attempted to distinguish between the functions of sociology and those of established disciplines such as philosophy and historical materialism. Efforts to extend the authority of sociology into these fields assumed an integrative rather than colonial character as sociologists asserted their prerogative over empirical work, which was implicit within but not addressed by other more 'theoretical' disciplines. At the same time, other new fields such as social psychology and scientific management were vying for this same privilege within the academy. Sociologists therefore also drew distinctions between themselves and other newcomers, claiming that sociology was not only empirical, but also 'scientific' and holistic and therefore deserving of exclusive authority over the empirical study of social life. This assertion of exclusive authority, however, was also contingent upon the public image of sociology, in particular the discipline's claim that it was able to transcend existential influences such as politics and personal prejudice in the pursuit of truth about social reality, while being pragmatically relevant for legitimate uses of power. Efforts to distinguish sociology from existing disciplines and assert its superiority over competitors were therefore combined with attempts to establish its 'distance' from all illegitimate power relations, as well as from 'dilettantes' and 'pseudo-sociologists' who presented threats to its fledgling legitimacy. Similar patterns of attempts to simultaneously expand, monopolize and protect the autonomy of sociology have remained consistent in the post-independence period, and have intertwined with other forms of boundary-work in sociology, such as the delineation between 'indigenous' and 'Western' knowledge.

The socially situated nature of sociological theory means, however, that we must be critical of the limits and possibilities of using theories of boundary-work, developed and applied primarily in the Anglo-European world, to understand the politics of knowledge in Central Asia. One point of tension relates to definition of the 'field' itself. Many of Bourdieu's original insights are useful here, particularly, the use of field as an analytical construct to theorize symbolic struggle in specific arenas of intellectual and professional activity. The field must still, as Bourdieu argues, be understood as a 'locus of a competitive struggle' for scientific authority, legitimacy and resources (Swartz 1997). However, the texture of this varies widely, and in particular, 'the structure of the intellectual field of the social sciences varies considerably across nations [as they] have their roots in the specific intellectual, institutional, and political constellations under which "social scientists" have tried to develop discursive understandings of their societies' (Wagner and Wittrock 1990: 6).

The field of sociology in Kyrgyzstan, for example, extends well beyond the borders of scientific institutions and includes sites of political, economic and industrial power, as well as symbolic domains of collective identity. This may be generalized, albeit cautiously, to other social scientific fields in other Central Asian republics. In the Soviet academy, social scientific authority was not exclusively an academic matter. Knowledge production was functionally integrated with social, economic and political production in other institutions, particularly technology, industry, the Communist Party and education. Many scholars were members of the Communist Party and scientific work was considered part of an individual's political activity; the title *obschestvoved–kommunist* [social scientist–communist] reflects this politico-intellectual identity (Tabyshalieva 1983). A brief report on the 'first conference of Kirgiz sociologists' published in Russia's major sociological journal *Sotsiologicheskie issledovaniia* [*Sociological Research*] in 1990 mapped the broad 'field' of sociology at the time. Present at the meeting were not only academic sociologists employed in universities, schools, factories and political organizations, but also representatives of industry, Soviet government officials, high-ranking members of the Communist Party and schoolteachers (Isaev and Niyazov 1990). In other words, while the concept of 'field' is heuristically helpful for expanding the imagination about where certain forms of knowledge such as 'sociology' might actually be located in a society, it must be employed creatively to take into account the different field structures in different societies.

Similarly, the definition of a professional 'market' must be expanded to include symbolic as well as economic capital; further research should be conducted on the extent to which Soviet professions and intellectual life could be considered 'competitive markets' in any sense. Research in Soviet Central Asia was highly centralized, with political organizations (sometimes the Communist Party, but also local industrial committees, agricultural unions and social organizations) making formal decisions regarding funding and organization. Competition for 'symbolic capital' such as professional prestige had a different meaning than it does in less centralized science systems where academics compete for individual grants, research ratings and the like. In Soviet Kirgizia, for example, the institutions responsible for bestowing legitimacy and granting resources were located at a geopolitical distance in the scientific and political centres of Moscow and Leningrad. The dynamics of boundary-work in sociology thus qualify Bourdieu's assertion that scientific authority necessarily 'owes its specificity to the fact that the producers tend to have no possible clients other than their competitors' (1975: 23) and that internal competition is more significant than external demand. In both Soviet and post-Soviet periods, reliance on external support (political, industrial and more recently commercial) has been a vital factor in the institutionalization of social scientific work (see also Graham 1998 for consideration of the scientific field in Soviet science more generally).

Overall, boundary-work is a useful concept for analysing the history and politics of Central Asian social science, which has been historically shaped by recurring debates over what constitutes 'good' and 'relevant' knowledge, what reliable methodology looks like, whether social research should be funded and used by the state and/or foreign organizations, how social science disciplines are related to one another (e.g. sociology, historical materialism and ethnology) and social practices (like market research and public opinion studies), and above all what relationship it is to have with politics (i.e. society's ruling 'power structures') and capital. Sociologists in Kyrgyzstan have continuously sought to expand their legitimacy, monopolize the right to construct legitimate images of society, and establish themselves as suppliers of social information for the purposes of political decision-making, economic progress, social discipline and human development.

Embracing the study of power and knowledge within the scientific field at these levels instead of seeking to escape it, and shifting from the particular ideology critique of competing truth-claims in sociology to a more holistic analysis of the institutional, intellectual and political fields that these claims are embedded within need not, as is often feared in Central Asia, signify surrender to the distortions of power and harmful subjective influences. Nor is it a negation or rejection of social science. It is better understood as a reflexive method for intellectual empowerment; as a way to reflect upon the partiality of privileged knowledges that often masquerade as universal and objective scientific truths. In the words of Mannheim (1936: 47), 'relativism and scepticism compel self-criticism and self-control, and lead to a new conception of objectivity'. And if we believe Bourdieu (1988: xii), this 'sociological critique of sociological reasoning' may in fact be a driving force behind the development of the discipline.

Knowledge and 'globalization'

Questions emerging from the sociology of knowledge, tied up as they are with theories of state and society, must also be contextualized within the history of colonialism, decolonization and global capitalism, particularly with regard to the conditions under which post-Soviet states have become integrated (or not) into global markets (Kandiyoti 2002; Rumer 2000). Former Soviet republics began nationalizing within a web of contradictory economic, political and cultural processes which were well under way. When not denoted as 'the transition', this is often simplistically referred to as 'globalization', but it has specific forms taking root in postsocialist space – to wit, new foreign policies of 'democracy promotion', capitalist triumphalism and ideological hegemony, aid dependency and the rise of indigenous authoritarianisms (Bryant and Mokryzycki 1994; Robinson 1996; see Kellner 2002 and Sklair 1995 for more on globalization). The terms of engagement with these processes have had marked and mixed effects on the way in which knowledge is organized and imagined in Central Asia.

Much has been made of the relationship between 'globalization' and social science. The sociology of globalization has to some extent eclipsed world-systems and dependency approaches to the study of global capitalism, and early projects to 'nationalize' and 'internationalize' social science are being rethought in terms of intellectual globalism (Schott 1993; Smart 1994; Therborn 2000) or cosmopolitan science (Beck 2000, 2002). Old divisions between 'Western' and 'Third World' academics are in certain ways diminishing (Arjomand 2000; Schott 1998; Smart 1994). The increasing interconnectedness of the world or at least the acute sense of this from within dominant systems – has disrupted older modernist categories of social analysis, particularly the nation-state unit which frames so much of contemporary sociological theory, and the multifarious presences which have emerged from formerly peripheral societies onto the international stage have challenged reductivist usages of 'centre' and 'periphery'. Most importantly, the presence of a global consciousness in specific localities creates entirely new forms of human experience and, in some societies, a new kind of sociological subject which is irreducible either to local or global analysis: the 'cosmopolitan subject' (Beck 2002).

A major part of this process is the struggle that often emerges in the act of being 'global' from a 'local' position (or indeed vice versa), and few would argue that cosmopolitanism is yet, or even should be, a reality. Excepting the most vulgar evaluations of 'globalization', there is no doubt that the transformation of social life in a period of late capitalist modernity is fraught with economic inequality and political and epistemological tension. There is also no illusion that global consciousness is necessarily associated with postmodern consciousness. For example, Beck argues on the one hand that we must dismiss the narrative of progress 'as spelled out in old modernization theory and theories of development [which] locates the non-West as the far end of an escalator rising toward the West, which is at the pinnacle of modernity in terms of capitalistic development, secularization, culture and democratic state formations' and that we 'need to attend to how places in the non-West differently plan and envision the particular combinations of culture, capital and nation–state, rather than assume that they are immature versions of some Western prototype'. On the other hand, he points out that these visions are often imaginations of 'alternative modernities' that are 'constructed by political and social elites who appropriate Western knowledges and *represent* them as truth-claims about their own country' (2002: 22, italics in original).

One of the most interesting aspects of intellectual life in post-Soviet Central Asia is its paradoxical relationship to these 'alternative modernities'. The representation of modernity being crafted, particularly as regards science and scientific knowledge, is less 'alternative' than 'returnative'. It aims to reclaim a previously hegemonic Western ideal of capitalist modernity which was denied by the imposition of a Russian socialist one, and which is viewed as a prerequisite to achieving both national autonomy and global recognition. The 'postmodern' is seen as two steps into the future of historical stages; the post-postmodern, unthinkable. Ironically, some of the elements of local modernity which become visible through the lens of a cosmopolitan as opposed to monological or universalizing consciousness are in fact what Beck identifies as threats to this very condition epistemological essentialism, nationalism, ideologies of global capitalism and 'democratic authoritarianism' (Beck 2002).

Cosmopolitanism is a critical alternative to traditional capitalist and culturally imperialist theories of 'globalization', as well as an attempt to move beyond the epistemological dichotomies of global/local and self/other that constrain thinking about global capital, politics and society, and knowledge. However, it is vital that we interrogate the discrepancy which exists between such anti-essentialist theories of society and knowledge, and the deliberately national, modernist and essentializing forms of organized social knowledge which are being articulated in the most isolated and yet simultaneously globally engaged localities in the world. This can be best accomplished by understanding the historical encounter of Soviet and Western modernities in indigenous experiences of colonialism, decolonization and recolonization, and by clarifying the meaning of globalization in the region. The globalization of knowledge in and of Central Asia does not currently take cosmopolitan form, but manifests in more familiar postcolonial patterns, many of which were considered extinct following the disappearance of the 'second world'. When studying the academy in Central Asia, however, it becomes clear that in the post-Soviet era, new globalist relations of domination and deference,[4] psychological states of inferiority and superiority, unequal and conditional access to material or intellectual resources, and structural and cultural dependencies have emerged to replace or compound those developed during Russian rule. While some can be attributed to the influence of neoliberal development policies these too can be considered postcolonial phenomena (Gosovic 2000; Liu 2003; Rist 1996). Any analysis of the academy in Central Asia is therefore incomplete without some familiarity of the sociology of knowledge in colonial and postcolonial societies.

2

THE COLONIAL LOGIC OF
SCIENCE IN CENTRAL ASIA

The dissolution of the Soviet Union engendered in the Central Asian republics diverse processes of decolonization and recolonization (Egorov 2002; Henderson and Robinson 1997). The partial unravelling of relations of dependency established over a century and a half of Russian rule and the emergence of new dependencies intended to fill these gaps, along with concurrent projects of nationalization, internationalization, cultural hybridization and essentialization, situate Central Asian societies firmly within the postcolonial world. However, there has been some debate as to whether Central Asia may be considered legitimately 'colonial' (Abdurakhimova 2002; Caroe 1953; Clem 1992; Fierman 1990; Tillett 1964) or 'postcolonial', particularly as this term is generally reserved for what was once known as the 'Third World' or more recently the 'South' (Beissinger and Young 2002; Cavenaugh 2001; Gammer 2000; Kandiyoti 2002; Moore 2001; Verdery 2002). I suggest, with Cavenaugh (2001) and Moore (2001), that complications of this labelling notwithstanding, serious comparative studies of Central Asia and the more classical postcolonial regions of Africa, the Middle East, India and Latin America are both necessary and fruitful. In many ways, as Moore argues, 'East is South' (2001: 115).

This is certainly true in the realm of knowledge institutions, specifically the social sciences. The debates occurring about sociology in Central Asia have local specificities and consequences, but they are not unique to the Soviet or post-Soviet condition. Since the mid-twentieth century, the politics of social scientific knowledge have been prominent on the intellectual landscape of newly independent states worldwide (Eisemon 1982; Fernandes 1967; Wagner and Wittrock 1990). There are a number of reasons for this, all of which are relevant to the analysis of organized knowledge in contemporary Central Asia, where colonial relations have historically shaped the emergence and development of social science.

First, social science is often perceived as a practical instrument for the realization of large-scale utopian reforms and planning strategies that accompany decolonization and national independence (Gendzier 1985; Hulme and Turner 1990). The form this takes depends on how notions of social reform are constituted, as well as on variations in schools of social scientific thought. In many of the anti-colonial movements of the 1960–80s, for example, politicized scholars produced critiques of intellectual dependency and developed indigenous theories of colonialism, underdevelopment and alternative approaches to national development (one of the most prominent being Frantz Fanon; see also Ake 1982; Clinard and Elder 1967; Joshi 1995). In Central Asia there has to date been a different trajectory, whereby the intellectual foundations of Marxist philosophy have been refunctioned to ground a 'new type of administratively-oriented knowledge' or scientific politics oriented towards the creation of a modern, capitalist nation-state and as a scientific 'corrective' to the manipulation of definitions of social reality by power elites (Bekturganov *et al*. 1994; Isaev 1995; Isaev *et al*. 1994d; Ismailova 1995; Migration 1992). Critiques of global capitalism and liberal ideologies of democracy, long considered a hallmark of postcolonial writing, have little resonance in these new cases where the colonial past was itself 'socialist'. Nevertheless, the close, seemingly organic association of social science with economic and social reform in newly independent societies naturalizes what is in fact a very complex relationship between social scientific knowledge and political power.

This relationship is further complicated by the fact that social science frequently assumes symbolic value as an indicator of modernization and Westernization or, conversely, nationalization and indigenization (Pertierra 1997). The development of sociology in post-Soviet societies accordingly cannot be divorced from its broader associations with both Western modernity and national sovereignty and identity. For example, in some cases sociologists make claims of universal scientific authority, represented by 'Western' science, to justify particular research methods or topics. Random sampling techniques are defined not only as 'objective', but also apolitical and democratic. This characterization of quantitative method is, in turn, associated with stylized images of 'Western democracy' and contrasted to methods used by sociologists in what are defined as traditional, 'unscientific' and backward societies (Baibosunov 1993). In a symbolic example, in referring to each other's theories, methods and professional ethos as either 'European' or *aksakal*-like [a word meaning 'elder' and generally referring to the patriarchs of traditional society], members of the Kyrgyzstani social scientific community reinforce symbolic parameters for the geopolitical scope of sociological knowledge in Kyrgyzstan (see, for example, Bakir Uluu 1994; Ryskulov 1998). The symbolic meaning of science, in other words, is flexible and used strategically within the postcolonial context; it is also intimately related to the more deep-seated dichotomy between 'European' and 'native' which has been both transformed and strengthened since independence from Russian rule (Khalid 1998: 72).

Finally, as an institution endowed with responsibility for defining social reality, social science can be a site for elaborating and negotiating competing theories of society and a platform for translating these theories into practical programmes for social change (Ake 1982; Bourdieu 1975; Bujra 1994; Eades and Schwaller 1991; Gosovic 2000). This, of course, depends significantly on the kind of relationships which are forged between academic social science and the state, the market, or 'civil society', and which are often challenged during processes of decolonization. The quest for scientific authority is highly visible in Central Asian academic and media institutions, where vigorous debates about the definition and role of sociology have been intertwined with discourses about national independence, revival and possibilities for development. For example, discussions of content, methods and professional ethics in Kyrgyzstani sociology are tied to deeper concerns about the fate of truth and role of social scientific knowledge in a society which believes it can and should restore the progressive promise of a scientific truth that has been hijacked by illegitimate power, not once, but continually.

The sociology of knowledge in colonial and postcolonial societies places particular emphasis on these sorts of questions, where the relationship between power and knowledge is often made explicit at every level, from individual intellectual thought to the structural organization and management of science as an institution. It often differs from the mainstream sociology of knowledge in the Western industrial societies that constitute the core of the scientific world system (Filino 1990; Schott 1992) and situates general theories about the ordinary or inherent politics of knowledge in the context of more critical theories: the role of organized knowledge in colonial domination and movements for liberation, the relationship between 'indigenous' and 'colonial' or 'national' and 'Western' knowledges, the dynamics of cultural or intellectual imperialism, the psychology of colonial and colonized intellectuals, the effects of structural and cultural dependency within academic communities, and processes of assimilation, resistance, alienation and hybridization. In other words, while not abandoning the insight from Bourdieu that social scientific knowledge is inherently heteronymous by its very nature, the difference in a colonial context is that such knowledge is subordinated to political and economic agendas for purposes of coercion, incorporation, control or exploitation, by members of proxies of a dominant power in institutions supported and maintained in a peripheral location to that dominant power.

There are obviously vast differences between Russian colonialism, Soviet imperialism and Western development and neocolonialism, and it is inaccurate to group them together under one conceptual umbrella. However, there are some unrecognized and sometimes denied consistencies in their *modus operandi* and consequences. All draw strength from materialist and evolutionary theories of modernization and are linked to the expansionist and civilizing missions of imperial powers. All have been to varying degrees grounded in Orientalist beliefs that non-Western societies are 'unscientific' and therefore 'backward', and purvey what Fanon argued was a central feature of colonialism; that

> by a kind of perverted logic, it turns to the past of the oppressed people, and distorts, disfigures, and destroys it. The work of devaluing pre-colonial history takes on a dialectical significance today.... Colonialism has made the same effort in to plant deep in the minds of the native population the idea that before the advent of colonialism their history was one which was dominated by barbarism.
>
> (1963: 169)

Although this is not always an intentional one and may be more of a 'total' than a 'particular' ideology, is based primarily on deep-seated assumptions of cultural superiority and inferiority, not only between 'Europeans' (or North Americans) and 'natives', but also in a more general 'clash-of-cultures' sense. It is also not necessarily externally imposed, but, as in the case of Central Asia, may be internally generated – indeed, this was one of the intentions of *korenizatsiia* (Liber 1993). During the eighteenth and nineteenth centuries, Russian colonizers laid grounds for the creation of academic and scientific institutions in Central Asia amidst virulent debate, both within the Russian community and between Russians and Central Asian intellectuals, over the necessity, morality and even possibility of leading cultural reform and 'modernization' in the region. Russian or 'European' culture was exalted over tribalism, nomadism and illiteracy, although it was not yet a policy to eradicate the latter. During the twentieth century, as Russian colonization gave way to Soviet imperialism, establishing scientific research and higher educational institutions in Central Asia became a prominent element of the 'Sovietization' of the new socialist republics. Soviet development was discursively anti-colonial, and by the 1940s either tolerated or encouraged hybrid Soviet–national and Marxist–Muslim identities among Central Asian intellectuals (Lewis 1986). However, it aimed to eradicate even those 'cultural survivals' that had been allowed to exist under tsarist influence. In the twenty-first century, these policies and practices themselves have become defined as undesirable survivals from the Soviet past – 'barbaric' not because of their affiliation with a tribal Orient, but because of their connection to totalitarian communism. As Noam Chomsky has argued, 'we are now proceeding to uplift the people liberated from communism as we've in the past liberated Haitians and Brazilians and Filipinos and Native Americans and African slaves and so on' (2000: 39).

In order to understand the contemporary politics of knowledge in Central Asia – the structure of the academy, the activities that do and do not go on within it, the struggles and successes of social scientists, the epistemological debates, the role of organized knowledge in social change – we must first understand the history of its emergence. For here, one of the most basic philosophical paradoxes of social science, the drive for scientificity in a field of knowledge that is inherently political, is embedded within wider histories of modernization, colonization and globalization. How does this affect the actual meaning and role of social scientific knowledge in the region, particularly among academics? It is to this that we now turn.

Science, the very idea

Science matters in Central Asia. The ideal of usable scientific truth – of rational, objective and legitimate knowledge produced and employed to further human, social and economic development – lies at the history of modernist development projects in the region, from tsarist colonialism to post-Soviet democratization;

from Sovietization to nationalization. Evidence about its actual effects, however, is uneven and contested. This is particularly true of the social sciences. While it has been argued that social science is underdeveloped and has played little role in the region's history (Egorov 2002; Gleason and Buck 1993), some indigenous academics maintain that their work has been central to economic, technological, agricultural, industrial and social development (Blum 1990; Ismailov 1995; Tishin 1980, 1998). Both assertions are exaggerated. The first is premised on an ethnocentric philosophy of knowledge, which operates with a highly normative definition of 'science', denies the possibility of any legitimate or relevant knowledge being produced under conditions of authoritarianism or state socialism, and privileges evidence of the overt politicization of social science as an ideological prop under Soviet rule. While it might be argued that science on the whole was inefficiently organized (Beliaev and Butorin 1982; Egorov 2002), to believe that, as one ethnologist put it, 'there is nothing there to study' would be to ignore the decades-long struggle to understand how social scientific knowledge should be defined in relation to Marxist philosophies of science, the Soviet centre and 'the West'. It also negates the subjective definitions of scientific truth created by academics attempting to negotiate the logics of science and power, and obscures the ways that both are sometimes subverted in the mundane processes of creating knowledge.

The second assumption, that social science has played a fundamental and progressive role in political, social and economic life in Central Asia is also overstated. It is more accurate to say that under the Soviet and to a lesser extent the post-Soviet regimes, academics working in the social sciences collected and interpreted information that contributed to the everyday management and planning of industrial, social and political institutions under Soviet control. More theoretical research, including challenges to these institutions themselves, was censored (often by researchers themselves). The rhetoric of a science-led society is thus largely based on a mythologized professional identity which has been cultivated as part of academics' struggle to gain legitimacy and create demand for their profession in an environment which demands they assume simultaneous and often competing roles of intellectual autonomy and social engagement.

It is not unusual to observe these competing images of Soviet and post-Soviet social science as being either entirely dysfunctional and politicized or superior and highly efficacious – this is in fact a common feature in the 'public understanding of science' more generally (Collins and Pinch 1993). Verifying the actual 'impact' of social science on Soviet and post-Soviet Central Asian societies and ascertaining how successfully social scientists fulfil their professional raison d'être are largely hermeneutical questions. Definitions of both legitimacy and relevance are historically contingent. What is ascertainable, however, is that these definitions have been closely tied to the economic, political and cultural agendas of colonial powers in Central Asia, and that the cultural meaning of social science has been shaped by the encounter with colonial practices. The idea of science is located in a web of identities, beliefs and social relationships related to colonization and nation-building, which pits 'modern' against 'traditional' and

'West' against 'east'. The failure or success of scientific institutions has historically been linked, often through Orientalist and Occidentalist discourses, to the social fate of Central Asians themselves – their 'adequacy' vis-à-vis modern nation-states, the degree of their agency or dependency in political affairs and the meaning of their acceptance of or resistance to various forms of 'Westernization'. Such dichotomous ideologies, perhaps predictably, appear in Russian literature about the 'development' of Central Asian societies during the early and mid-twentieth century (Sahni 1997). However, they are also commonplace within the discourses of contemporary Central Asian social scientists who employ them either in earnest or as rhetorical devices to increase their own authority by distancing themselves from stigmatized ways of life and modes of thinking.

While it is important to understand the political and institutional practices that have shaped the politics of scientific knowledge in Central Asia, we must therefore also consider how the cultural meanings of knowledge influence these practices. Colonialism is of course not the only cultural logic that mediates knowledge production in Central Asia. But it is an extremely important one, affecting the development of social science, its epistemological foundations, the organization of teaching and research, and the professional identities of Central Asian academics. Theories of culture and colonialism are therefore vital for understanding how and why a new modernist science project has emerged in Central Asian sociology and what consequences this might have for intellectual practice today.

Culture and colonialism

Theories of 'cultural colonialism' (sometimes referred to as 'cultural imperialism', though these obviously have distinct implications) achieved controversial notoriety during the 1960s, emerging as radical postcolonial alternatives to more economically or politically based theories of imperialism that had been dominant in political science and 'area studies'. Social scientists in newly independent nations, as well as Western critics of counter-insurgency or anti-communist and later neoliberal development policies, were among the most vociferous employers of the approach, although it is now often stigmatized as 'anachronistic' (Prashad 1998: 23). The limitations of this perspective, particularly regarding the fluid boundaries of both 'culture' and 'colonialism' and the haziness of interchanging concepts of 'colonialism' and 'imperialism' have thus been clearly and usefully articulated in recent years (Tomlinson 1991). We must therefore be particularly conscious to move beyond a negative critique of the conditions of colonialism towards recognizing, where it exists, the limitations of its reach, the localization and hybridization of its practices, and the complicated effects that it has on cultural production.

The more specific concepts of intellectual, scientific and academic colonialism describe the conditions and political forces shaping the uneven development of academic disciplines and scientific research in postcolonial societies. They were embedded in wider critiques of the triumphalism of 'scientific development' and of a growing tendency at this time to import 'Western scientific solutions' to abstract or non-existent social problems in the Third World. Even before Johan Galtung (1971) formalized the idea in his 'structural theory of imperialism', sociologists had used the term 'academic imperialism' in descriptive and sometimes polemical work about the development of 'Third-World' social science (Fernandes 1967: 330; Whyte 1969). A volume of increasingly theoretical and analytical work about the topic later developed in India (Rahman 1983) and Africa (Ake 1982) throughout the 1980s, with an emergent concern about the globalization of American sociology (Lamy 1976).

Interest in intellectual or academic colonialism had waned by the end of the 1980s. However, there is still considerable work being written on the topic. Some theorists, recalling Shils' (1970, 1988) theory of 'academic centrality' in international social science, talk about the 'intellectual domination' of the West and the asymmetric or unequal relations between postcolonial scholars and their American and Western European colleagues (Bujra 1994), sometimes referring to the development of a 'scientific world system' (Schott 1998). Others focus on more locally situated power relations between 'local' and 'foreign' scholars. Some writers have attempted to redefine academic colonization in more discursive ways, or even critically *as* a discourse (Tomlinson 1991) and to frame it in more Gramscian terms within contemporary global politics as a new 'global intellectual hegemony' (Gosovic 2000). There have been a number of studies on the relationship between academic knowledge, development practice, and globalization and/or imperialism in 'developing' countries (Cooper and Packard 1997; Escobar 1991; Long and Long 1992; Watts 1993). Some of the most recent work in cultural colonialism comes from Alatas, whose definitions of 'intellectual colonialism' and 'academic dependency' are particularly useful for our purposes.[1] According to Alatas (2003: 600), intellectual colonialism is defined as the 'cultivation and application of various disciplines such as history, linguistics, geography, economics, sociology and anthropology in the colonies' to bolster the 'control and management of the colonized'. The 'colonial mode of knowledge production' is characterized by six elements: 'exploitation, tutelage, conformity, secondary role of dominated intellectuals and scholars, a rationalization of the civilizing mission, and the inferior talent of scholars from the home country specializing in studies of the colony' (Alatas 2003: 601). As sociology in Kyrgyzstan exhibited each of these characteristics during the Soviet period, this concept is a useful conceptualization of its political economy of knowledge.

Academic dependency, the child of intellectual colonization and a condition that develops during or as a result of decolonization, can be defined as a 'condition in which the social sciences of certain countries are conditioned by the development and growth of the social science of other countries to which the former is subjected'. In the case of Soviet science, we must speak of 'republics' as well as 'countries'. It develops when one social science community becomes dependent on 'the institutions and ideas of Western social science such that research agendas, the definition of problem areas, methods of research and standards of excellence are determined by or borrowed from [another]' (Alatas 2000: 603; see also Lamy 1976). Alatas (2003) identifies six characteristics of academic dependency: dependence on externally produced theories and ideas, on the foreign media of ideas (such as books, journals and conferences proceedings), on foreign educational technologies, aid for research and education, foreign investment in education, and demand for skills in the West, or brain drain. These dependencies often evolve after decolonization, when the colonial mode of knowledge production shifts to a postcolonial or neocolonial model.

In fact, the colonial logic of science in Central Asia became most visible during this more recent phase. The dissolution of the Soviet Union had profound implications for the organization of social science and reconceptualization of its role in post-Soviet society. In Central Asia, this had two main consequences: the breakdown of centralized economic and political relationships between science and the state, and the decolonization of the Central Asian periphery from the Russian centre. The loss of state funding and subsidies, partially replaced by investment from foreign governments and development organizations, laid ground for certain features of academic dependency. Cultural decolonization, in the post-Soviet context known as de-Marxification or de-ideologicization, occurred in tandem with the influx of new ideas from 'the West' and created conditions for new intellectual dependencies to emerge, including recolonization by anti-communist forces.

Despite the wealth of theoretical and empirical material available on the subject, discussions about cultural colonialism are almost absent from the discourse of development in post-Soviet Central Asia. Concepts of 'imperialism' and 'colonialism' are often rejected by both Western and Central Asian scholars who interpret them as derogatory labels as well as, or perhaps more than, analytical categories. They rub against historical memories of Soviet Central Asia's identity as a progressive model for decolonizing nations during the mid-twentieth century (Ali 1964; Altbach 1971; Nove and Newth 1967; Salehi-Esfahani and Thornton 1998). Indeed, the debate is politically and emotionally charged, and if relations of dependency are not also explored in terms of resistance, negotiation and agency, they serve only to reify the power relations that already exist. However, the concepts do help theorize the structural inequalities, ideological discourses and power relations that shape both the social and political arrangements of organized knowledge and localized professional practices in the region.[2]

Knowledge, 'civilization' and control in Central Asia

The intimate relationship between knowledge, power and identity in Central Asia was evident even before Soviet reformers began creating scientific institutions in the borderland republics during the early twentieth century. Harnessing the authority of modern knowledge and the advantages of industrial technology (particularly in printing and publishing) was a central concern for the Jadids,[3] Muslim reformers who advocated the 'modernization' (albeit not the secularization) of

37

thought and education in Central Asia as a method for combating Russian domination in the region (Khalid 1998). Then as in the future, ideologies of enlightenment, European-led progress and modernity were embraced, engaged, adopted and adapted amongst Central Asian intellectuals both as elements of the colonial experience and as forms of resistance to it. After the October Revolution and the replacement of tsarist with Soviet administrators in the region, the cultural authority of Islamic knowledge and Jadidi education was gradually discredited and displaced by the secular developmentalism endorsed by local Soviets. The intellectual traditions of Muslim scholars, which had intersected with secular values and technologies of the European enlightenment to create a unique alternative modernity in the region, were silenced under the new regime. However, the belief in modern, rational and scientific knowledge remained.

Science and scientific knowledge also occupied a central place in early Soviet Central Asian society. Early Bolshevik programmes for 'civilizing' the Central Asian steppe and incorporating its diverse tribal communities into a new Russian empire included the creation of universities and research centres in the region. Modernist philosophies of both *obrazovanie* [education] and *vospitanie* [upbringing] were central to Soviet programmes of cultural reconstruction: new Soviet citizens could be moulded and disciplined here; class, ethnic and gender inequalities ostensibly redressed; individuals socialised into Russian history and culture. As Russian policy in Turkistan shifted from watchful tolerance of local Muslim authorities to the permanent institutionalization of colonial power, Soviet authorities were particularly eager to counter pan-Turkic and pan-Islamic movements, which they feared might result in the emergence of a culturally legitimated political threat (Panarin 1994; Ro'i 1995).

Because the use of brute military force to abort the establishment of tribal and religious alliances in the region had historically met with violent resistance, imperial power was increasingly exercised through the control of cultural and intellectual life in Central Asia.[4] Modernist theories of knowledge were integrated into larger campaigns to secularize and sedentarize the semi-nomadic tribes, which had been forcibly settled through collectivization during the 1930s; they were also pitted against more fluid traditions of scholarship in the intellectual centres of what had become Uzbekistan and Tajikistan. In Alatas' words, ideologies of science were used to 'rationalize the civilising mission' – development and social prosperity would be brought to the region through the creation of modern scientific institutions, which would in turn contribute to the region's progress. In this context, status was bestowed on fields of knowledge and disciplines that would enable Central Asian economies to fulfil specialized functions in the Soviet system of national production and distribution. This was not only exploitative, but also gave local scholars 'secondary' roles in Soviet science, channelling their ideas and productivity into the needs of empire. Finally, Soviet authorities aimed to minimize and stifle dissent by bringing indigenous elites into the folds of metropolitan power through tutelage and the *korenizatsiia* [indigenization, or assignment of native elites to key posts] of academic and scientific life.

During the 1920s and 1930s, however, the agenda in Central Asia was the development of basic literacy, not the institutionalization or indigenization of social science; Soviet schools existed side by side with more traditional educational institutions such as the *maktab* and *madrassa* until the 1930s (Khalid 1998; Nove and Newth 1967; Smith *et al.* 1998). Although institutions of higher learning such as universities and filials of the Academy of Science began to appear in Kirgizia in the 1930s (Karakeev 1974) following the establishment of the Central Asian University in Tashkent in 1920 (Ali 1964; Simirenko 1969a), these were primarily oriented towards political and technical education, not teaching or academic research. However, early educational and scientific institutions had functions beyond their use as experimental sites for promoting literacy and disseminating public pedagogies on science, politics and morality. They were also 'bases' for Russian ethnographic and cartographic research.

During the late nineteenth and early twentieth centuries, social science in Central Asia consisted primarily of Russian-led colonial 'expeditions' to Turkistan, as the region was then referred to. Research trips were organized to gather information about the languages, customs, religious beliefs, productive capacities and political structures of the various ethnic groups living in the region, for the purposes of their more effective incorporation within the Russian, and later Soviet, empire (Khalid 1998; Tabyshaliev 1984). During the 1920–40s, they also became bases for anthropological expeditions to local communities, which were oriented towards exploring the new empire's natural and human resources, and which grounded and justified early policies regarding 'nationality' in the region (Kul'turnoe stroitel'stvo 1974: 21; Tabyshaliev 1984). The research centres that were set up in the Kirgiz Autonomous Oblast in the 1920s, such as the Academic Centre and the Scientific Commission for the Oblast Branch of People's Enlightenment, were organized to support Russian-led research within and about the area, particularly its natural resource potential (Tabyshaliev 1984). At the same time, the Central Asian Bureau, Central Asian Economic Council and Central Asian Territorial Commission (1924) remained subordinate to the Russian state and Communist Party (Wheeler 1966: 67). During this period, Russian ethnographers and sociologists coordinated research groups and institutes in the region in order to understand its potential as a source of natural and labour resources (Karakeev 1974: 14, 17).

The frequency and scale of field expeditions to Kirgizstan and other Central Asian regions increased during the 1930s as part of the effort to 'construct Soviet culture' in the area. In 1935, the USSR Central Scientific Commission issued a statement that such work 'allow[ed] for the significant and thorough illumination and clarification of the fundamental problems facing the national economy of Kirgizia' (Kul'turnoe stroitel'stvo 1974: 21). Towards the end of this decade, the president of the USSR Academy of Science and member of the Committee on Filials and Bases argued that 'the Kirgiz Republic can no longer fulfil premises from the centre of various types of expeditions [if the] work is not attached to constantly operational filials of the Academy of Science in the regions'.[5]

The Kirgiz filial of the USSR Academy of Science was established several years later in 1943. While this was publicized as a major scientific development, it also symbolized Russia's increasingly penetrative imperial ambitions during the post-war period. Shortly thereafter, localized institutions were transformed into republican Academics of Science, a move that had ideological as well as professional and economic implications for social scientists.

The Russian-led development of scientific research in Kirgizia entailed more than mere data gathering; in fact, it pervaded the very theoretical foundations of social science in and about the region. For example, the debate to clarify where Central Asian pastoral-nomadic societies belonged in the Marxist five-stage categorization of social evolution was perceived by some Russian scholars as central to the 'socialist reconstruction' of the 'Soviet East' (Gellner 1988: 99). The publication of such debates in regional newspapers and journals such as *Sovietskaia etnografiia* [*Soviet Ethnography*] contributed to the creation of a body of knowledge about Central Asian societies within the Russian and broader Soviet academic community. Tsarist-era research on Central Asian societies that focused on Islam and shamanism, traditional family structures, tribal kinship relations, patriarchy and indigenous folkways was used both to expand understanding and to justify and administer 'modernization' in Central Asia. It was transformed into data that informed efforts to integrate the region economically and politically while simultaneously promoting 'national traditions' in social and cultural life (Park 1976: 6).

Centre and periphery in Soviet social science

Central Asian social scientists have frequently adopted Russo-centric narratives of the history of Soviet science, positing the 'Great October Socialist Revolution' as the precondition for their own intellectual history. As one Kyrgyz scholar argued in the 1980s, for example, the 'straight scientific and systematic study of Kirgizia began only during Soviet rule, when, among other socialist transformations, the culture of revolution was realized in the periphery' (Tabyshaliev 1984: 162). The development of Soviet social science about Central Asia, along with the establishment of educational and scientific institutions in the region, was until recently interpreted in as the first stage of enlightenment. In Kirgizia the opening of the Academy of Sciences in 1954 was celebrated as a 'historical event in the life of the Kirgiz people, bearing witness to the growth of its economy, science and culture' (Ob uchrezhdenii 1962). Russian experiences were universalized into general 'Soviet' ones, and many academics still refer to earlier periods of repression in the Russian Soviet Socialist Federated Socialist Republic (RSFSR) when explaining the underdevelopment of the academy in the republics (Baibosunov 1998; Isaev 1998c; Ryskulov 1998).

The early history of sociology in Soviet Russia is of course an important part of its history in Central Asia. The institutions in which sociology emerged in the region during the 1960s – the university, factory, Academy of Sciences and

Communist Party – were integral parts of the state's politico-industrial-ideological apparatus (Balázs *et al.* 1995). Social scientists were directly responsible to Soviet and party authorities at the local, republican and all-union levels; they were dependent both structurally and culturally on institutional centres in Moscow, Leningrad and Novosibirsk; on organizations such as the *Komsomol*, and on industry for funding, resources and professional opportunity. In addition, they were in many ways doubly subordinate: ideological and political deference to the Soviet state and Communist Party, requisite even for academics working in the centre, was compounded by the fact that for Central Asians this was also an ethnocolonizing power. Intellectual and material inequalities were not only embedded in the organization of social scientific knowledge, but influenced how academic knowledge was produced, interpreted and legitimated in the republics. In order to understand the historical development of sociology on the imperial periphery, therefore, it is also important to understand how it was related to the organization and politics of sociology in the Soviet centre. As there are already a number of excellent studies of the history of Soviet sociology (Beliaev and Butorin 1982; Ivanov and Osipov 1989; Matthews and Jones 1978; Myrskaia 1991; Shalin 1978; Simirenko 1966; Weinberg 1974, 2004; Zaslavsky 1977; Zestov 1985), I will concentrate on those aspects which are significant for understanding the history and politics of sociology in Central Asia.

Until the early 1920s, Russian sociology flourished as a young discipline in the new Soviet political system. The Russian Sociological Society was founded in 1916, a sociology department was opened in the new Institute of Psycho-Neurology in St. Petersburg, the People's Commissariat on Education approved the establishment of the Petrograd Socio-Bibliological Institute in 1918, and translations of European social theorists such as Spencer, Comte, Durkheim, Weber and Simmel were made available in addition to the works of Marx and Engels. At this early stage of his political career, even Lenin believed that sociology might be instrumental in theorizing relations between Russia and the newly incorporated 'nationalities', and in 1918 he founded a new Socialist (later Communist) Academy of Social Sciences to 'make a series of social investigations one of its primary tasks' (Matthews and Jones 1978: 3; see also Batygin and Deviatko 1994; Urban and Lebed 1971). For a brief period, 'the Bolsheviks seem to have believed, like their tsarist predecessors, that [sociology] favoured their cause' (Matthews and Jones 1978: 4; see also Shalin 1990).

Within a few years, however, fears that 'bourgeois' sociological theory would threaten the legitimacy of the fledgling political regime led to a sudden change in policy regarding scientific and intellectual life. These marked what Alex Simirenko (1969a: 6) called the 'period of decline' in Soviet sociology. By 1922, the party had banned the teaching of sociology in universities and closed both the Socio-Bibliological Institute and the Russian Sociological Society. Prominent sociologists such as Pitirim Sorokin (perhaps better known as founder of Harvard University's Sociology Department), whose work had become influential under the more intellectually liberal conditions of the early Bolshevik regime, were gradually suppressed by the party's increasingly authoritarian control over the academy. Sorokin left Russia that same year to escape the persecution that many of his colleagues had already been subjected to (Simirenko 1969a). While empirical research continued on demographics, working conditions, family relations, the effects of propaganda and time budgets until the mid-1930s, the authorities' need to minimize exposure of the empirical brutality and logistical shortcomings of Stalinist policies put an end even to these narrowly defined studies.

According to Matthews and Jones (1978: 4), the need for sociology, as well as its political viability, ended in 1936 when Stalin announced that 'society, having achieved "socialism," now consisted of two friendly and internally homogenous classes (the workers and the collectivized peasantry) with a "stratum" or *prosloika* of intelligentsia drawn from both classes but having no contact with the means of production itself'. There was no longer a need for the academic study of society; only for its explanation and illumination through the lens of historical materialism, soon to become synonymous with scientific sociology (Weinberg 2004). The small corps of sociologists practising in the RSFSR and other republics was liquidated through dismissal, exile or execution during the 1930s. The official conflation of 'sociology' with Marxist–Leninist theories of social laws and historical development marked the beginning of a long struggle to delineate the boundary between sociology and politics in the Soviet Union. Marxism's status as the foundation of all scientific knowledge, as well as its colonization of other social sciences, laid the ground for future controversies about the definition and role of sociology in the region.

This generalized history of Soviet sociology, however, obscures the fact that neither periods of creativity nor repression in the early history of the social sciences in the RSFSR were reproduced in Central Asia. In fact, the republics as we currently know them did not exist when the seminal struggles for power ensued in Russia, and remained at the time in conditions established under Russian colonial rule. In 1924, the Central Asian republics were only just being carved into existence by the Communist Party in its *razmezhevanie* [demarcation] of administrative boundaries for the new Soviet state (Gammer 2000: 128; Tchoroev 2002). 'Elder brother' narratives about the impact of Russian scientific knowledge on local intellectual life, however, became dominant in so far as 'the idea that Russian colonialism was more progressive than the British and other colonial enterprises finally came to dominate Soviet historiography', including the history of science and intellectual life (Tchoroev 2002: 360; see also Ali 1964: 92; Critchlow 1972). There is a further explanation for the prevalence of this interpretation, which is that for many Soviet scholars there was no subjective discrepancy between generalized and local experience (Nesvetailov 1995: 66). Even in the periphery, 'Soviet science perceived itself as being the center' of world science. Central Asian scholars, many of whom identified personally and professionally as Soviet citizens as well as members of a 'nationality' or republic on the imperial periphery have therefore long rejected critical appraisals of this early period of social science in the region.[6] They have been particularly resistant to its definition as 'colonial', asserting, for example, that 'the development of the social sciences in the USSR is a single, total process, to which scholars from the Central Asian republics have contributed.... We do not have "central" and "peripheral" science, but a single Marxist–Leninist science about society' (Leninizm i razvitie 1970: 30).

Such sentiments remained commonplace in sociological texts well into the 1980s; there was no 'rebellion in the academy' against Russian domination in sociology as there was in the discipline of history during the 1960s and 1970s (Allworth 1998: 72). While it is necessary to acknowledge the significance of this identity for many scholars working in remote regions of the empire, it is also important to recognize that the history of Soviet social science was neither singular nor unproblematically progressive. The anti-imperialist rhetoric of a union of equal nations bore little resemblance to the colonial empire itself, which was based on a very strong and deliberately maintained differentiation between centre and periphery, in the organization of science as much as in other social institutions; indeed, in the mind.

Republican sociology: the case of Kirgizia

While the form and content of republican and Russian sociology were similar, the conditions within which sociology emerged in the periphery and centre differed. Social scientists in Central Asia had even less intellectual freedom and fewer chances for occupational mobility than their counterparts working in the Russian centre (Critchlow 1972: 23). This imbalance was not lost on one sociologist, later a professor of anthropology at the American University–Central Asia, who draws clear distinctions between Kyrgyzstani and Russian-led research in Kirgizia prior to the 1960s:

> There had been some investigations organized by Russian sociologists and they went to Issyk-Kul Lake [the country's largest alpine lake]. There were ethnographic and sociological investigations of... rural life and relations between people, family and marriage relations, social relations and ethnographic relations of the inhabitants of the villages of Chichkhan and Darkhan.... But these were done by Russian sociologists. And then our Kyrgyz sociologists were also involved in such investigations.
>
> (Asanova 2003)

Such relations of dominance and dependence are common where there are divisions between centre and periphery. Contrary to assertions that there was neither centre nor periphery in Soviet science, a number of scholars have recently argued that the centre–periphery relationship was in fact one of its most prominent characteristics (Eisenstadt 1992; Nesvetailov 1995; Schott 1992). Sociologists themselves increasingly support this. One sociologist and academics administrator at a Kyrgyzstani university, for example, argues that one of the main reasons for the 'crisis' in post-independence sociology is that 'science in Kyrgyzstan did not develop independently. Its financial base, structures, themes, theories, etc. were all directed from Moscow; thus, when the Soviet Union collapsed, the entire science structure collapsed with it'.

Kirgizstani academics' early relationships with the Soviet centre were ones of deference and dependence, which manifested itself in both structural and cultural forms. Developments in social science in Moscow and Leningrad during the 1950s and 1960s affected the way sociology could be defined and practised in the periphery. Central Asian academics looked to Moscow to learn how to affiliate themselves with and/or distance themselves from other social science disciplines, organize scientific activities, and relate to industrial and educational institutions and the broader field of political power, as well as to ascertain what constituted legitimate 'sociological' problems to be studied. In some cases, the centre came in the form of Russian academics sent to establish scientific institutions in the republic. Until the 1980s, social scientists emphasized the constructive role of Russian assistance in the advancement of indigenous scholarship, claiming that 'the process of the formation and development of the Kirgizstani intelligentsia occurred through the brotherly assistance of the Russian people: many scholars worked in Kirgizia, helping to establish national cadres' or 'thanks to the emergency assistance of the Russian people and the Leninist nationality policies carried out by the Communist Party of Kirgizia, it became possible for scientific workers – social scientists – to grow' (Alimova 1984: 36–37). The dependence of early Kirgizstani social science on Russian material and intellectual support is accurately acknowledged in these narratives. However, until late perestroika there was little critique of the political causes and cultural consequences of this dependency. For example, the 'national cadres' or indigenous elites produced through these efforts were often Moscow-oriented throughout their careers, and in the absence of stable indigenous sociological institutions they were dispersed to work as 'individual enthusiasts' (Isaev 1991b), labouring in isolation or in small teams.

In other cases, Kirgiz social scientists travelled to the centre. In the 1980s, for example, the number of candidate and doctoral degrees in the social sciences increased primarily as a result of educational exchanges in which Kirgizstani students were educated in Moscow and Leningrad, and Russian scholars travelled to Kirgizia for 'consultations' (Skripkina 1983: 17).[7] However, such exchanges did not result in the institutionalization of sustainable republican institutions primarily because they were fundamentally unequal. Institutionally, social science in Kirgizia developed according to definitions of social science that were developed in the Russian centre.

Orientalism and Occidentalism

This inheritance was legitimized not only by the professional ambitions of Kirgizstani sociologists who sought recognition from prestigious academic authorities in the Soviet centre, but also by entrenched attitudes of Orientalism and Occidentalism within the academy. A sense of inadequacy *vis-à-vis* the centre on the part of 'Eastern' scholars and of superiority in the opposite direction was cultivated through years of Russification in the region. The construction of non-Russian ways of knowing as inferior was intertwined with the construction of Marxist–Leninist science and Soviet rationalism as superior; 'orientalisms are created out of a dialectic that also produces occidentalisms' (Restivo and Loughlin 2000: 139).

In his treatise on the hegemonic othering of non-Western societies, Edward Said defined Orientalism as a discourse that is used to construct and maintain colonial relationships and politico-cultural divisions between the 'East' and the 'West'. The 'essence of Orientalism', according to Said, is an ideology of 'ineradicable distinction between Western superiority and Oriental inferiority' (1978: 42) in the organized knowledge about 'Eastern' societies (i.e. Middle Eastern, African and Asian) produced within the context of the European colonial project from the eighteenth to the twentieth centuries in the discipline of Oriental Studies. The field was organized around the assumption that there is a clear distinction between 'us' (Europe, the West, familiar) and 'them' (the Orient, strange, exotic); of 'self' and 'other'. The result was the production of a rigidly dichotomous discourse which purported that 'on the one hand there are Westerners, and on the other there are Arab–Orientals; the former are (in no particular order) rational, peaceful, liberal, logical, capable of holding real values, without natural suspicion; the latter are none of these things' (Said 1978: 49). This, Said argued, not historical evolution, is how the 'West' emerged as dominant and the 'Orient' as in need of colonization (1978: 38, 49).

Although Said's theory was elaborated on the example of the British and, to some extent, Western European academies, it is also an apt description of the imagination of Central Asia in much Russian social science of the period (Allworth 1975: xxx–xxxi; Borozdin 1929). Even in Soviet-era literature, Central Asians were often represented as tribal, nomadic, traditional, dirty, ignorant and backward, a threat to modern culture and civilization, and in need of guidance and enlightenment (Panarin 1994: 63) – as *ostalye narody* or 'backward people' (Cavenaugh 2001: 12). Within this milieu, ethnographic and sociological studies about the region were designed to catalogue and reify differences between Russian and Central Asian cultures and aid in 'civilizing' the latter through Russification. This not only enabled the more effective administration of the Central Asian region, but also contributed to the creation of a hierarchy of region, ethnicity and language that later became embedded in the Soviet academy. Normative and naturalized differentiations between the 'East' and the 'West' led to the institutionalization of a form of scientific racism which extended beyond the boundaries of Russian academic elites into the scientific disciplines and into the collective consciousness of Central Asian social scientists themselves (Cavenaugh 2001). Pronouncements of *bratstvo* and *ravenstvo* [fraternity and equality] notwithstanding, the Orientalist and Occidentalist foundations of social science in Central Asia were not eradicated by the Bolshevik revolution, the evolution of the Russian empire into the Soviet Union, or indeed by national independence. Racist and colonial representations of Central Asia were further institutionalized in 'anti-imperialist' and pro-development Marxist critiques of tsarist politics.

The assimilation of the history of social science in Central Asia into the grand narrative of the Soviet 'civilization' of Central Asia, the use of Soviet science as a yardstick of social development and national pride in Kyrgyzstan, historical amnesia about the repression of alternative historical narratives, and post-Soviet counter-reactions to each of these tendencies may be seen as responses to the colonial logic of social science in Central Asia. The history of sociology in Kyrgyzstan must therefore be analysed within the broader framework of the Russian colonization of Central Asia and its continuation in the politics of Sovietization.

It has been argued, however, that the logic of power in former Soviet space is never totalizing, and that there are always spaces of agency in cultural work (Lubrano 1993; Verdery 2002; Yurchak 1997). To what extent has this been true of social scientists working in the non-Russian republics of the Soviet Union? Did members of the intelligentsia – scientists, academics, teachers and political activists – develop hybrid identities working within the formal structures of the academy? What did Soviet social science look like on the periphery? The experiences of academics in Central Asia suggest that the centralized organization and vertical control of scientific activity in the Soviet Union did not have the totalizing effect on knowledge production that it is often assumed to have had. Sociologists, for example, interpreted and engaged with the Soviet science system in various ways. They were aware of its inherent inequalities and those who found this problematic struggled to redress them through the strategic use of state subsidies and creative interpretations of Soviet ideology. In addition, many academics that did not fully subscribe to the technocratic model of state science were professional divided between loyalty to state and society, on the one hand, and to the quest for scientific truth on the other. In order to understand more about the actual practice of Soviet science in Central Asia, we can view its history through a postcolonial lens, focusing on how it was experienced, interpreted and constructed on the periphery.

3

THE SOCIAL SCIENCE PROJECT IN SOVIET KIRGIZIA

The origins of sociology in Kyrgyzstan – when, where and why the field came into being as a 'field of knowledge, academic discipline and professional practice' (Isaev 1999a) – is largely a hermeneutical question which depends on how sociology itself is defined. Founding narratives have therefore become matters of contention amongst Kyrgyzstani sociologists since the republic's independence as rival histories of the field's development emerged in the construction of a new disciplinary identity and competition in a new professional market. Some, predominately those trained by or associated with the late Asanbek Tabaldiev, claim that his sociological laboratory established within the Department of Philosophy and Historical Materialism at the Kirgiz State University (KSU, later the Kyrgyz National University or KNU) in 1966 was the first sociological laboratory in the republic and that therefore he should be considered the 'founding father' of sociology. Those affiliated with Kusein Isaev argue that his laboratory, opened in the Department of Scientific Communism at the Frunze Polytechnic Institute (FPI) in 1983, was the first 'national' sociologist and that he therefore deserves to be called 'father of Kyrgyz sociology'. Others, including Isaev himself and outside observers and younger scholars, argue that sociology had been stagnant from the 1960s (Blum 1993) and that opportunities for its development only emerged during perestroika (Isaev 1998b; Isaev *et al.* 1994b).

Thus far, there has been no concerted effort to analyse how these narratives have been constructed, what they represent, or what they reveal about the intellectual and social forces shaping the institutionalization of the discipline; no attempt to place the narratives in a broader historical context that would shed light on how the epistemological and institutional legacies of Soviet sociology have influenced the development of the field in the post-independence period. These narratives therefore remain at the level of first-order experience, what Mannheim (1991: 50) referred to as 'immanent interpretations'. Both imagine scientific knowledge as developing in a linear, internally lawful and progressive way and categorize alternative epistemologies as deviations from the 'correct' historical trajectory. They are thus often presented as mutually exclusive and competitive: each are accepted by their advocates as an accurate portrayal of historical reality over and against more ideologically or politically motivated alternatives.

47

However, sociologically, we must understand the webs of meaning within which these interpretations have emerged and the political processes through which they are construed as significant. Whereas traditional disciplinary histories such as those being currently constructed in Kyrgyzstan often tend to simplify the narrative and make it coherent, it is also useful to explain why opposing 'legitimate' narratives exist in the first place (Thompson Klein 1996: 205).

In part, founding narratives of sociology in Kyrgyzstan are rhetorical weapons in a struggle for scarce material and cultural resources that is experienced throughout the post-Soviet academy (Huskey 2004). Competition for students became fierce in universities when educational policy shifted from a socialized model towards a new 'market model' of education. In the initial absence of clear standards in this context, many Soviet-era lyceums and technical colleges reclassified themselves as universities and began recruiting students for new 'marketable' programmes such as business economics, marketing and accounting.[1] In universities subordinate to the Ministry of Education, as well as in those answering to private boards of trustees, social science departments are increasingly forced to demonstrate their viability by increasing their student intake. This has thus engendered 'competition between the [major universities] regarding the preparation of specialists, and each one strives to show that they have had the best developments in sociology, that they have the best professors' (Omuraliev 2003). Some academics feel that this task is complicated by the devaluation of social science on the whole; that, as in the humanities, students – and female students in particular – 'either think they can't work as scientist or they don't believe it is worth trying as they see so few results from the scientific community at the moment. Instead, they move into more profitable areas of work like business' (Asanbekov 2003).

In this highly competitive environment, the establishment of authority became a central part of projects to institutionalize the field within the universities. Clarifying an institution's historical identity is one way to establish public confidence. However, because the authority of one institution is often enhanced by the de-legitimization of another, institutional identity construction has become rooted in competition. As Bekturganov *et al.* (1994) remarked, the hope of the discipline lies in sociologists trained 'at two parallel institutions by only a few scattered professionals' and 'this separation... contributes little to the creation and development of sociology, and to the preparation of cadres for this prospective branch of knowledge'.

While symbolic conflicts over which institution may be considered the 'original' source of sociology in the republic are partially motivated by material needs and professional interests, they are grounded in epistemological debates about whether the institutions in question actually conducted 'sociological' work during the Soviet period, whether their theories and methods should be considered legitimate contributions to the discipline as it is currently defined, and whether their work is relevant to contemporary Kyrgyzstani society. The term 'sociology' does not necessarily signify the same set of ideas and practices across time and space. Juxtaposing alternative narratives on the origins of sociology in Kyrgyzstan illustrates how its meaning has fluctuated throughout the history of the field.

Sociology as renaissance and invention

Soviet histories of Kirgizstani social science assert that 'the beginning of the democratization of society at the end of the 1950s and beginning of the 1960s facilitated the *ozhivlenie* [revitalization] of the social sciences.... Researchers in the republic began to develop new problems, lay down new scientific orientations and strengthen the connection between philosophy and practice' (Kakeev 1990: 38; see also Simirenko 1969; Yanowitch 1989). Indeed, sociology first emerged as a field of knowledge in Kirgizia during this 'renaissance' of Soviet sociology, which was itself part of broader reforms in Communist Party ideology and organization (Remington 1988).[2] The rapid growth of sociology 'offices' in academic and industrial institutions was not linked specifically to the rationalization of social engineering, and though it did not mark the birth of an autonomous discipline, it signalled an 'evolution' in the intellectual atmosphere of the times, which eventually facilitated the emergence of sociology (Beliaev and Butorin 1982: 429; Hollander 1978: 375; Shalin 1990: 1019). It was considered novel because it seemed in sharp contrast to the severe and systematic repression of sociological research during previous decades, when, as Weinberg argues, 'sociology as an independent academic discipline virtually disappeared in the Soviet Union [and] MarxismLeninismStalinism took its place' (1974: 8). Because the controlled reintroduction of sociology in the post-Stalin period was integrated into de-Stalinization and intended to give a socio-cultural face to economic development and industrialization, it did not represent a return to greater academic or intellectual freedom. However, it did provide a more pragmatic impetus for the development of institutions in which social scientists would begin to pursue these and other goals.

The reforms of this period had ideological as well as administrative components, and the 'new' sociology was implicated in both. Politicians and social scientists alike began to assert that the 'Marxist science of society' should play a key role in the scientific development of socialist society (Simirenko 1969a). While the Communist Party is infamous for its repression of critical sociology during this period, the development of an infrastructure for empirical sociological research was in fact a party-driven process which revolved primarily around the need for strategic information about public opinion on party activities during a period of rapid industrialization, and for 'scientific proof' that its reforms were successful (Matthews and Jones 1978; Tabyshaliev 1984; Tabyshalieva 1986). Sociology, then defined as empirical research conducted by historians, philosophers, economists and psychologists, was redefined as a 'scientific, objective, comprehensive approach to social problems' (Simirenko 1969: 392). The need to give the Communist Party an ideological makeover and create a veneer of communication between authorities and the public gave rise to increasing interest in public opinion studies. 'Connected to information offices in many party committees [were] opinion polling services, nearly always using amateur sociologists. ... Breakdowns of occupation, age, party status, and such data [were] intended to help party speakers tailor their addresses more closely to their audiences' in workshops and factories (Remington 1988: 62).

The centralized organization of science meant that the emergence of sociology in each of the republics was driven by and responded to many of the same forces that compelled its re-emergence in the RSFSR: industrialization, rationalization, secularization, political and economic administration and bureaucratization, and re-ideologicization. As one mathematician who trained as a sociologist in Kirgizia's first sociological laboratory in the late 1960s remembers,

> [a]t the end of the 1960s and beginning of the 1970s there was powerful support to develop all spheres of life in Kyrgyzstan. In economics, they built new factories, production plants and organizations...and as it was in the economy, so it was in culture.... It was during this very time, during the ascendance of Kirgizstan, that sociology came into existence.
>
> Tishin (2003)

Isaev, too, recalls that 'this was a time of enthusiasm for sociology: positions for "sociologists" or "social psychologists" were made in industries, organizations, even *kolkhozes* and *sovkhozes* [collective and state farms]' (Isaev 1991a; Tishin 1998).

As demands for information about discrete questions concerning social stratification, urbanization, occupational prestige and leisure and the family became more pronounced, empirical or 'concrete' sociology enjoyed a rapid revival in the Soviet Union (Lubrano 1977: 38). However, it occupied an awkward position *vis-à-vis* the ideological orthodoxy, as empirical research had long been dismissed as 'unscientific', 'bourgeois' and dangerous. Dominant theories of knowledge during this period were anti-empirical and 'Marxist–Leninist'; empirical data were deemed irrelevant or invalid (Greenfeld 1988: 109; Lubrano 1977). Thus, as sociolgoists' interest in empirical research increased throughout the 1960s and 1970s, there were 'some major conflicts over the nature of the discipline, or more accurately, over how one approaches sociological enquiry in the context of Marxism–Leninism' (Lubrano 1977: 37). This provided the conditions for boundary-work in the field, which delineated 'good' from 'bad' sociology in the context of epistemological doctrine.

The resurgence of empirical research during the 1960s thus represented more than a change in party policy toward sociology. It also implied changes in the interpretation of Marxist theories of society and its relation to social engineering. As it became clear that theoretical formulae were inadequate predictors of problems or tensions within Soviet society, empirical research was re-branded as a 'science of prognosis' that could be used to forecast things such as 'demographic and ethnic processes, urban development, the social effects of scientific and technical progress, changes in the social pattern of society and developments in public education, health and culture (Mandel 1969: 57). This type of research did not contradict the grand theory of Marxist–Leninist philosophy. Rather, it was seen as a technical supplement, useful for solving practical problems as opposed to generating theories of society (Hahn 1977: 38). Theorizing and empirical research, while interrelated, were considered autonomous practices that informed but did not intervene in one other (Hahn 1977; Rumiantsev and Osipov 1968). As will be discussed later, this formulation neither compelled nor satisfied many social scientists, and in Central Asia led to repeated attempts to rectify these two tasks.

The loosening of restrictions on empirical study led to a boom in academic and public interest in sociology during the 1960s. Ultimately, much of the data gathered about Soviet society in this period was deemed threatening to Brezhnev's regime, and sociologists endured a new wave of repression during the 1970s (Brym 1990: 207; Weinberg 1992: 2–3). The 1960s, however, were years of growth. They also marked the beginning of sociology in Soviet Kirgizia.

Tabaldiev and the 'first group of students willing to be sociologists'

It was against this backdrop that in 1966, Asanbek Tabaldiev organized a sociological laboratory attached to the Department of Philosophy and Historical Materialism at KSU. It had a variety of purposes: educational, ideological, professional and *vospitatel'nye* [training or upbringing], reflecting the formal integration of education, professional training, scientific research and industrial production in Soviet society. It provided space for informal research training that was not otherwise available in the official structures of science during this period. Formal teaching of sociology was prohibited at this time, and there was widespread concern among sociologists throughout the USSR that this threatened their professional status.[3] The expansion of 'laboratories', which were formed on a voluntary basis by trusted academicians and which required little financial or administrative commitment from the state, was meant to ameliorate the situation.

Initially, as the laboratory did not challenge the party's intellectual orthodoxy, it developed in relative autonomy from and in relative conformity to central academic authorities. It was subsumed within a department of historical materialism, supervised by a respected member of the republican intelligentsia, constituted as an extradisciplinary space for uniting students and teachers interested in empirical research and oriented towards the pragmatic study of problems in 'communist construction' such as industrial management and inter-ethnic relations. Informal study groups on social theory were organized within the relatively flexible purview of Tabaldiev's own ideological work, which, given his political standing he had some autonomy to oversee. The little institutional security Kirgizstani sociologists enjoyed in the confines of the KSU laboratory was revoked, however, when members of the group later began to associate normative values of patriotic social service and scientific objectivity with 'good' sociology and corrupt political opportunism and manipulation of empirical data with 'bad'. By challenging the political truth-claims of 'Marxism–Leninism' with empirically based sociological ones and presenting research results that suggested alternative and even critical answers to rhetorical questions posed by the Communist Party about the state of ethnic relations in the republic, the laboratory was stripped of its status as a legitimate scientific institution.

Nevertheless, for many social scientists that were trained during the 1960s and 1970s, this laboratory is the earliest institutional point of reference in the history of sociology in Kyrgyzstan. One sociologist still working at the same university claims that the establishment of this laboratory marked the beginning of more than 30 years of sociological research in Kyrgyzstan (Tishin 1998), and another attributes to Tabaldiev the creation of an 'entire school on the problems of the theory of nations and national relations' (Omuraliev 1997).[4] Many of Tabaldiev's former students thus consider him the founder of sociology in Kyrgyzstan, a title that has become highly contested in debates about whether Marxist sociology, or indeed any academic work produced during the Soviet period, should be included in the contemporary history of science and ideas in Kyrgyzstan (Bekturganov 2003).

Tabaldiev wore many hats during his life. He was a scholar, village school-teacher, head of the Department of Marxist–Leninist Philosophy at KSU, and secretary of the university's *Komsomol* committee (Smanbaev 1986). A philosopher by education, he aimed to elevate the intellectual and professional level of social research in the Kirgiz Republic, organizing the laboratory to combine training in both sociological theory and research and setting rigorous standards of academic conduct for his associates. He invited students and young teachers to attend a biweekly *nauchnyi krug* [discussion group], where they discussed topics ranging from the classical works of Marx and Lenin to new publications in social philosophy and sociology. These meetings seem to have extended beyond their ideological purpose and provided spaces of creativity in an otherwise moribund intellectual culture. One participant recalled,

> you cannot say working under Tabaldiev was easy: he gave much time, spirit and thought to his work and demanded the same from us.

> The discussions were heated and time flew by unnoticed, as in all this there were elements of play and humour.... The meetings began at 3:00 in the afternoon, after the end of lessons, and sometimes lasted until 9:00 or 10:00 at night.

> (Nurova 2000: 14, 15)

The composition of this 'club' as it was sometimes referred to (Asanova 2003) reflected the interdisciplinary, or perhaps more precisely extradisciplinary, status of sociology at the time. The field was diffuse in its early years, conceived as a specialized methodology as opposed to a discrete discipline (Simirenko 1969). While some students considered themselves 'sociologists' even at this early stage, they hailed from a range of subjects, including philosophy, history, historical materialism, political economy, linguistics and mathematics. Umut Asanova (2003), for example, studied English, 'I was very much interested in sociological investigations', she says, 'and took part in those which were organized by the department... because in Kirgizia there had been no preparation for teachers of philosophy, [and] they were all integrated from different specializations.' Similarly, Nurova wrote a doctoral dissertation on the development of *malykh narodov* [minorities] under socialism. 'It was in effect about ethnosociology', she says, 'But at that time this type of committee did not exist, and I therefore defended it as historical materialism and received a degree as a Candidate of Philosophical Science in Social Philosophy'. She asserts that she had personally always distinguished between sociology and social philosophy, 'My dissertation was on sociology when sociology did not have its own status. But when it gained its own status, I became the first doctor of sociology' (Nurova 2003).[5]

In other words, for these women, sociological research was neither allied with nor positioned in antagonism to other social sciences. However, the association of the term 'sociology' with only empirical or 'concrete' research (i.e. both empirical and applied), left little space to carve out any sort of theoretical niche within existing academic disciplines in the Soviet academy. By the time the KSU laboratory was established, the boundaries of sociology had already been fixed in both political doctrine and philosophies of science. Previously, and predominately in the RSFSR, there had been considerable struggles to negotiate the boundaries between sociology and politics, 'Soviet sociology' and 'bourgeois sociology', and empirical sociological research and Marxist theories of society. Once sociology was subsumed within state ideology and segregated from competing theoretical traditions, however, such debates were reduced to superficial discussions about whether empirical research was or was not compatible with Marxist philosophy. Sociologists, most of whom were then self-educated researchers from other fields, attempted to mould sociology into these pre-existing parameters set by the political and ideological establishment in order to create space for themselves and their work within the dominant state and party apparatuses.

Industrial sociology in Kirgizia: therapeutic, disciplinarian and pedagogical

The emergence of *zavodskaia sotsiologiia* [factory or industrial sociology] in Soviet Kirgizia during the 1960s provides an excellent example of how the development of sociology was embedded equally within discourses of science, modernization and industrialization, and how modernist theories of scientific knowledge intersected with social technology and political discipline to shape the professional roles of sociologists. Former members of the KSU laboratory not only assert that studies of industrial management were sociologically interesting, but also that they were vital to industrial production and national development. One argued that

> from the middle of the 1960s, the results of social scientific research about the formation of agricultural plans were used to elaborate scientific bases for the accommodation of productive forces in the republic, to help economists help industries with production and increasing effectiveness.... In the 1970s, scholars essentially expanded the spectrum of research linked to the planning and prognosis of the socio-economic development of the republic. The analysis of regional particularities in economic development, revealing backward sections and disproportions in social production and the preparation of general recommendations from scientific research collectives, supported an increase in the level of management of the national economy in the Kirgiz SSR.
>
> (Tabyshalieva 1986: 329)

While the accuracy of such claims has been highly contested (Egorov 2002; Gleason and Buck 1993), such work did have other, more intellectual and professional consequences. As elsewhere throughout the USSR (Balázs *et al.* 1995), some of the first major sociological projects carried out in Kirgizia focused on issues of industrialization, particularly labour itself. Asanova, for example, has strong recollections of conducting research among workers at a time when the industry was expanding rapidly in the republic.[6] Her reflections reveal the culture of early sociological research, its hierarchical organization, and the interrelationship between political activity, industrialization and social science education:

> We went, if you can imagine the car... it was open in the back... there were perhaps twenty or thirty of us, students and teachers, and it was very cold on the way from Bishkek to Osh [Kyrgyzstan's northern capital and largest southern cities]... even in summer it was very cold in the heart of the mountains. We investigated the mining plant and the international, interethnic relations there, because it was a very great plant, and perhaps as many as fifty or eighty nationalities were represented.... We had questionnaires, of course prepared by the teachers.... Me, personally, I enjoyed it greatly – to ask, to observe and to write everything down. We were there for the whole month. And Asanbek Tabaldievich, he himself came and organized some meetings, [saying] please do it this way, interview that way. And of course we were, how should I say, perhaps inexpensive, even cheap working power for interviewing. But even so it was some sort of school for us, a school of how to behave with people while interviewing them.
>
> (Asanova 2003; see also Tishin 2000)

Through such experiences, students were socialized into relationships that shaped their broader sense of professional and social responsibility, their approach to social research, and their understandings of the relationship between knowledge and power.

This was also true in other sites of sociological work, such as the industrial sector itself. For example, researchers also worked with managers and party bureaucrats as part of a scientific disciplinary apparatus, united by a mutual agenda to promote 'cadre development' in industry. Their main role was to survey worker (dis)satisfaction and make recommendations for improving working conditions in order to minimize conflict and turnover in the local workforce, in the interest of the national economy (Omuraliev 2003; Tishin 1988). Some served on what were called 'social cadres committees' – known by workers as 'commissions for dismissals' – and either fired or assigned 'leaves of absence' to individuals deemed 'dysfunctional' for the collective goal (Sorokina 1989).

Official discourses articulated a particularly mechanical relationship between the empirical data obtained by sociologists, the application of their interpretations and social change. Knowledge about human behaviour could be applied as a technology for disciplining and modifying it (see, for example, Aldasheva and Nikolaenko 1973). However, while sociologists broadly embraced this role, there were also criticisms of the mechanization of the relationship between knowledge and practice, particularly in so far as it challenged their professional status and authority. In later appeals for sociology to be taken seriously by the Communist Party, for example, Tishin (1988: 63) issued a caveat, that 'the methods and results of sociological research [in industry] are not a panacea of all administrative "calamities," but merely a means that can be fully effective in the able arms of the administration'. This nevertheless affirmed the legitimacy of this administration, and issues such as inequality and power politics within the Soviet workplace were not considered valid subjects for sociological inquiry. In fact, as purely 'theoretical' questions, they were not considered sociological at all. The empirical study of social problems within factories was acceptable as long as it was conducted within the framework of Marxist–Leninist theory about the role, legitimacy and necessity of the institutional arrangements themselves. Discipline, satisfaction and indoctrination were not considered problems *sui generis*, but studied in terms of how their 'incorrect' development was an obstacle to economic efficiency and political conformity.

Research, in other words, served very 'therapeutic' ends. It was seen as a way to modify workers' behaviour and attitudes within the factory setting. This has been described elsewhere in critical theoretical critiques of this type of research, particularly as it was conducted in the US in the 1950s, insofar as

> [t]he therapeutic character of the operational concept shows forth most clearly where conceptual thought is methodically placed into the service of exploring and improving the existing social conditions, within the framework of the existing societal institutions – in industrial sociology, motivation research, marketing and public opinion studies.
>
> (Marcuse 1964: 107)

Nevertheless, technocratic policies of scientific management, particularly within industry, shaped the field of sociology for years to come. Sociologists also assumed a number of other duties within the industrial sphere as propaganda, agitation and political education were all elements of academic work.[7] As Omuraliev (2003) recalls, 'even in industry, they paid a lot of attention to what we call the sphere of *vospitanie*, and sociologists at that time were ideologically based'. Sociologists working in the Frunze Agricultural Machinery Construction Factory during the 1980s, for example, worked 'to influence, through all channels, the formation of a healthy moral–psychological climate' within the workplace and to 'improve communication' between workers. They installed a booth called 'Sociology into Production' within the factory and created programmes for the factory radio station and newspaper *Sel'mashevets* (*Agricultural Machine Producer*). 'By these channels', argued one, 'We propagandize sociological knowledge and familiarise the collectives with the results of studies that have been conducted' (Vlasova 1989: 42). Such activities were justified well into the 1980s by the belief that

[s]ociologists in industry are called on not only to study public opinion, but [to] fundamentally form it during the perestroika of society. Factory workers' participation in sociological research brings about the process of democratization; that is, their concrete participation in the adminis-tration of labour, production, living and the social life of the factory.

(Vlasova 1989: 43)

This was an era of management science and social engineering, when Stalin's expansionist modernization policies were being replaced with programme for increasing the efficiency and stability of the country's enormous military–socio-industrial complex (Beissinger 1988). These changes in the politico-institutional environment created new opportunities for social scientists to lobby for legiti-macy, respect and financial support from the Communist Party and Soviet state, which attempted through decree to channel the administrative powers of scientific research into Cold-War programmes of industrialization (Hahn 1977; Mandel 1969: 42).

Kyrgyzstani sociologists now tend to interpret this shift as evidence of increased recognition for their professional contributions to the improvement of socialist society. In certain ways, this is true, sociology was institutionalized in the republic in the context of this technocratic demand for empirical data on the scientific management of diverse populations and the ideological need to legitimize as 'scientifically true' increasingly invasive forms of social control.[8] Despite the production of voluminous rhetoric, however, sociologists received little de facto cooperation from industry and complained that only the Communist Party supported their work (Vlasova 1989). This, combined with increasing demands for sociologists to conduct research in the service of 'socialist construction', created considerable resentment, particularly as academic underperformance was sometimes attributed to 'national backwardness' within the Soviet Union or to totalitarianism from without (Chrissterson 1994). Tishin (1980), however, argued that sociology was stunted not because of anything inherent to republican culture, but because Kirgizstani students were 'ten to thirty years behind' those trained in Moscow with regard to the latest technologies in mathematical modelling and sampling (then popular research methods in industrial sociology). While political rhetoric advocating the new role of sociological research was quickly transmitted from the Russian centre to the peripheral republics, financial and intellectual capital for facilitating such developments was not forthcoming.

Nevertheless, industrial sociology continued to expand in Kirgizia during the 1970s. By the next decade, the republic boasted a number of large factories, some employing thousands of people, and an ethnically heterogeneous industrial workforce (Omuraliev 2003). Sociologists continued efforts to expand their 'services' throughout the industrial sector, arguing that factories were in need of 'highly qualified and experienced industrial sociologists' because 'without them it [was] impossible to create or carry out the types of modern plans of social development for collectives' (Tishin 1980). During perestroika, he continued to assert that 'great projects and plans [in Soviet industry] are often not realized or experience difficulties, not for technological and economic reasons, but because people do not want to work in programmes which do not take their interests, demands, social particularities and local traditions into consideration' (Tishin 1988: 63). Like many sociologists, he saw industrial sociology as a means of closing the cognitive gap between the objectives of centralized planning, the constraints of local conditions and the subjective experiences of individuals.

It was, after all, a logical niche for sociologists to fill during this period. By defining themselves as both 'scientific' and ideologically committed, they could stake professional claims for themselves within the party-led movement to rationalize Soviet industry. By the early 1980s they had, like sociologists elsewhere in the Soviet Union, 'established an identity as a specialist group in a position to legitimately influence social policy' (Hahn 1977: 34). This identity, however, was dislocated from the reality in which they remained subordinated to political decisions made by central authorities. Early members of the KSU laboratory confronted this tension most directly and particularly during their research on 'national relations' in Kirgizia, or as we might now understand it, studies of ethnicity, class, colonialism and identity.

Sociology and 'national relations' in Kirgizia

In addition to their work in industry, members of the KSU laboratory also emphasize the importance of their research on 'national relations' during the late 1960s and early 1970s (Tishin 2000), which, in seeking to make empirical evaluations of political rhetoric, challenged the officially sanctioned boundaries between Marxist science and bourgeois 'pseudoscience'. Tabaldiev's long-term interest in the study of 'the national question' in the Kirgiz Republic was not unusual; discussions of the 'nationalities' or different titular groups residing in the Soviet

Union, processes of 'internationalization' and the creation of a multi-national national identity were ubiquitous in Soviet social science at the time. While 'nationalities policy' had been on the social scientific agenda since the 1920s, Khrushchev's introduction of the 'sblizhenie-sliianie [rapprochement–merger] theory' in 1961 raised new questions about how ethnicity could and should be dealt with in social science. In Central Asia, where collective identities remained strong despite widespread Russification, the pronouncement of a 'new stage in the development of national relations in the USSR in which the nations will draw still closer together and their complete unity will be achieved' was met with some reservation (Rakowska-Harmstone 1972: 9).

Tabaldiev, who was particularly interested in how research into ethnicity might inform conflict prevention in Kirgizstan (Asanova 2003), was critical of how the issue was formally approached, namely, as a set of ideological tenets to be confirmed by academics as opposed to a question that could be empirically investigated. He resisted using research to legitimize what he considered anti-intellectual and inflexible elements in Soviet ideology, including the official party narrative of 'fraternity and equality among nations'. One student, for example, recalls that

> [w]hen I began my dissertation with citations from Khrushchev, he crossed them out and said that this was the work of subordinate party workers, but that for aspirants, scientific integrity is most important. If the citation was actually necessary, I could use it, but if it was used only to express loyalty, it had to be removed, which he did.
>
> (Nurova 2000: 18)

According to his students, Tabaldiev was sceptical of asking centrally produced and ideologically driven questions about social processes that he doubted were actually occurring, or occurring in particular ways, in the periphery. One argues that he was determined to understand whether 'there [was] *real* equality between nations, and how this . . . manifested itself in their lives, family life, street life, labour life, etc.' so that inequalities within Soviet society might be redressed (Asanova 2003).

Unlike most published work on ethnic relations in the non-Russian republics at the time, Tabaldiev's research challenged party doctrine by invoking earlier Leninist theories about 'dualistic tendencies' in group identification. This theory, which in other contexts was hailed as a contribution to Marxist theories of the state and thus Marxist sociology (*Kirgizskaia Entsiklopedia* 1982: 301), asserts that 'mature ethnic groups' of people encounter one in a dialectical relationship of assimilation and self-realization, with the tendency to liberate themselves from oppression as well as to submit to the 'historical values of the past ruling systems' (Rakowska-Harmstone 1972: 8). While Tabaldiev's interpretation of this theory affirmed the value of ethnic integration so celebrated in political rhetoric, it also allowed for the possibility that distinct republican identities might legitimately emerge, or indeed conflict. Thus, while there was nothing theoretically uncommon about the application of Leninist theory to Kirgiz society, contemporaries argue that this was nevertheless threatening to authorities who were at the time attempting to quell the rise of nationalist sentiment in the Central Asian republics and construct a universalized Soviet identity:

> This was his great investigation, seeing two directions of international development.... Now we see there really was a kind of assimilation, because people assimilated their language, their style of life, clothes, and their behaviour. *This is also a kind of invisible assimilation, when Kyrgyz people still began to speak only Russian, and just gradually forget their own language.* It is also a process of assimilation. So first [he investigated the] integration, how it occurred, *and the second, growing nevertheless – growing, with this parallel direction or tendency, the growing of self awareness. Who am I, what am I? Am I Kyrgyz, or who? And what am I doing as Kyrgyz?*
>
> (Asanova 2003, italics mine)

Tabaldiev published a doctoral dissertation on his dialectical theory of ethnicity in the Soviet Union in 1971. However, *Sotsializm i natsii* [*Socialism and Nations*], the large-scale empirical study of ethnic relations in Kirgizia carried out and written up collectively under the auspices of the laboratory, never saw the light of day. According to students who participated in this research, had the book been published it would have dealt a devastating blow to the official party line that there was a peaceful, progressive *sblizhenie* and *sliianie* of ethnic groups in the Soviet Central Asian republics (Nurova 2001; Tishin 2000). Instead, it is argued, it both revealed a strident Russification of the non-Russian population in Kirgizstan and exposed economic and social inequalities between Russian and non-Russian-speaking groups in the republic. It is also said to have documented a growing sense of ethnic consciousness among Kyrgyz communities, which was spawned not by 'ethnic narrow-mindedness' or 'chauvinism', as stressed by the Communist Part, but by colonialism, poverty and injustice.

Although Tabaldiev reportedly proposed amendments to nationality and language policies on the basis of this research, the results were first 'corrected' by party leaders; then banned from publication; and finally, as with subsequent studies of ethnic relations in the republic, they disappeared altogether. At the time, any work which challenged hegemonic claims that 'the national languages [were] progressing rapidly under the conditions of freedom and complete equality of nationalities in the USSR' and that 'every citizen of the USSR [was] guaranteed complete freedom to speak, educate and teach his children in any language whatsoever' was treated as deviant and seditious (Tadevosian 1963: 44). This was particularly true in questions about ethnicity, as was later revealed: 'in practice the sphere of nationality relations has been put beyond criticism, treated as a zone of general harmony, while anything that doesn't fit into that harmony is tossed aside, branded as a phenomenon of bourgeois nationalism' (Bagramov 1990: 74; Karklins 1986). Nurova, who worked on this project as an interviewer in the village of Chychkan (now Dzhenish), recalled why she felt the research was important and why she was so disappointed that it had been censored:

At the time of the survey I attempted to clarify from one young Kyrgyz woman, who did not speak in Russian, why she wanted to educate her children in a Russian school. She answered *'kyrgyzcha on klassty butkondon koro, oruscha uch klass oido turbauby'* [he finished tenth grade in Kyrgyz school, but it was as if he had finished Russian third grade[9]] and it became clear. She went to Przheval'sk with her son to visit her brother in the oblast hospital. They were there for half a day, but could not find out which section their relative was in. They were beset upon by workers in white coats, most of them Russian, who did not know (and who did not want to know) the Kyrgyz language. ... And in fact, without knowing the Russian language at that time, it was difficult to work and study, let alone climb the professional ladder, and no one demanded knowledge of the other state language ... particularly from Russians. In this situation Tabaldiev attempted ... to change national and particularly language policies in Kyrgyzstan at the core. He thought the thesis that Russian had become the second mother tongue of the Kyrgyz was untrue and argued against it. . . . But this point of view did not coin-cide with the officially declared approach, and thus many other prob-lems, including [those in] our book, were crushed. I always think about this with serious pain.

(2000: 16)

While Tabaldiev's interest in both ethnic relations and sociology was typical for social scientists of his generation, his combination of the two and desire to conduct empirical studies of ethnic difference and inequality in Soviet Central Asia were unusual and frowned upon by insecure and authoritarian party officials. His was a 'path untaken in the republic' and, eight years after he founded his sociological laboratory, Tabaldiev bowed to political pressures to abandon it (Tishin 1998: 31). The type of sociological knowledge developed here – critical of the status quo but deferential in its tone, working within the confines of Marxist–Leninist theory and loyal to the ideological spirit of the socialist project – was 'unscientific'. Criteria of legitimacy for social scientific truth-claims were articulated not by academics, but by the power structures they were embedded within. Alternative interpretations of social reality based either on social theory or empirical research could exist as long as they were kept private affairs, within the realm of the alter-reality that existed in parallel with Soviet officialdom.

The dissolution of the KSU laboratory

Some argue that such professional disappointments and intellectual frustrations contributed to Tabaldiev's death at 40 in 1975. Without speculating on this assertion, it is obvious from the historical record and interviews with his students that immediately after he defended his doctoral dissertation, the party began trying to persuade him to abandon the laboratory and sociological research. In 1973, he transferred to the Academy of Science, where he became chief editor of the party's *Kirgiz Soviet Encyclopaedia* (Nurova 2000). By some accounts, apparently excerpted from his personal journal, he was dedicated to this task and worked at it diligently until the day he died. He is purported to have written, 'who needs a person who has done nothing for his own people? I came to this work out of a precious honour and pride in my people, and therefore want to show that the Kirgiz are in no way worse than other groups' (Smanbaev 1986). While few of his former students would question whether this might be attributed to him, some take issue with official reports that his career move into higher party service was entirely a personal choice. Nurova (2000: 18), for example, has argued that her 'teacher was torn away from the school that he had built with ten years of his life and this, it seems to me, was the second blow that Asanbek-agai [as his students called him] did not survive'.

The rise and decline of Kyrgyzstan's KSU laboratory is portrayed either as an extremely marginal event in Soviet sociology or as a formative, almost mythical, period in the history of Central Asian sociology. Both are plausible, if incomplete, interpretations. But it is also a window onto the relationship between power and knowledge in early Central Asian sociology and onto the everyday experiences of academics working on the Soviet periphery. While members of the laboratory operated with a range of narrowly defined theoretical concepts such as labour, class and ideology, they also produced propositions about the social tensions which were brewing beneath the society's ideologically crafted veneer. Kuban Bekturganov, a philosopher who worked in Tabaldiev's laboratory and who is now an instructor of sociology at KNU, suggests that the studies simply exposed what was already tacitly known: 'What the party says, that's how it must be. But real life is different' (Bekturganov 2003). The exposure of the discrepancy between political ideology and social reality was not unique to Kirgizstan. It had been felt throughout Soviet society, particularly in the non-Russian regions and republics during the 1960s (Beliaev and Butorin 1982; Hahn 1977; Matthew and Jones 1978; Nove and Newth 1967). In Kirgizia, it manifested itself in the 'reorganization' of the KSU laboratory.

Tabaldiev's reassignment signalled the end of his sociological career. The laboratory survived the 1970s under the leadership of Rakhat Achylova, one of Tabaldiev's Leningrad-educated students (Asanova 2003). Others, led by Tishin, continued their research in industrial sociology. Even today, students of the 'Tabaldiev School' remain influenced by his re-interpretation of the social scientific terms which were previously used to classify ethnic groups in the Soviet Union, such as *narodnost'* [peoplehood or nationality], *natsional'noe men'shinstvo* [national minority], *natsional'naia gruppa* [national group], and *etnicheskaia gruppa* [ethnic group] (Nurova 2001). The Bishkek Humanitarian University Sociology Department holds an annual conference in his name, the *Tabaldievskye chtenii* [Tabaldiev Readings]. This tradition, however, has lost much of its historical and nearly all of its intellectual significance for late and post-Soviet generations of sociologists who are deliberately reoriented away from Soviet concepts of 'nation' and towards more 'western' notions of identity or nationally focused studies of 'etnos'. In addition, much of the material gathered and produced during the laboratory's eight years of existence was lost in its dissolution, its contributions to sociological knowledge in the republic relegated primarily to the personal memories and archives of these early researchers. As reflected in Asanova's (2003) regretful statement: 'there were a lot of... files, all these reports – but I don't know what happened. Where are the reports, all the reports we were reading?... I don't know where these reports are, I don't know'. Much that survives of this early laboratory is contained in the older generation's memories, images of a charismatic scholar whose life and death has come to symbolize sociology's struggle for existence in an authoritarian society.

First-generation sociologists made a fine distinction that later became blurred when Soviet-era science was denounced whole cloth as illegitimate propaganda. In their view, Tabaldiev was an effective propagandist and agitator, but he was not an ideologue. In other words, he engaged in what they defined as 'good' Soviet science, that which fulfilled the idealistic goals of socialist development, as opposed to reproducing the 'pseudoscience' of party bureaucrats who sacrificed scientific truth to pragmatic political power and who used the authority of fabricated scientific truth-claims to disguise potentially explosive tensions and inequalities. Younger academics, however, are sceptical about the reliability of this founding narrative and, as will be discussed, have begun to introduce alternative interpretations of this period. Tension between these narratives is again primarily interpretive, related to boundaries of 'the sociological' and of professional conceptions of appropriate relationships between knowledge and power, and past and future.

Social science on the post-Soviet periphery

At the time of his death, the Kirgizstani academic community acknowledged Tabaldiev's 'contribution to the history and theory of national relations and the training of scientific and teaching cadres', the importance of his many publications, his work in propagating 'political and scientific information among the masses' (Asanbek Tabaldiev 1975) and his status as a 'well-known specialist of dialectical and historical materialism' (Karakeev 1974: 86). However, the nature of these activities changed after Kyrgyzstan's independence, when he was designated the republic's first 'sociologist'. Since that time, the history of the KSU laboratory has been constructed among Tabaldiev's former students, most of which are now scattered throughout the republic's atomized departments, institutes and centres of sociology.

While in operation, the laboratory was neither entirely stifled by ideological politics nor able to offer a space for truly alternative thinking in the social sciences.

However, the attempt to combine empirical research with both theory and social policy made it a site of struggle between the authority of scientific knowledge and that of political power within Soviet society. Early studies that were banned were not acts of 'resistance' to official ideology, however tempting it might be to interpret them in this way. In their quest for professional legitimacy and social relevance, sociologists imagined the discipline as a thoroughly modern Soviet science. Their work is instead an excellent example of 'the paradoxical fact that great numbers of people living in socialism genuinely supported its fundamental values and ideas, although their everyday practices may appear duplicitous because they indeed routinely transgressed many norms and rules represented in that system's official ideology' (Yurchak 2003: 5). One of these values was scientific modernization. Kirgizstani sociologists were in many ways frustrated by their inability to be accepted as legitimate contributors to this project, both in the Russian centre and in their own society. For example, despite the fact that the Communist Party blocked publication of the laboratory's research on ethnic relations and only a small amount of work was ever published, Tishin (2003) still maintains that Soviet use and funding of sociology was superior to that of the current Kyrgyz government and foreign organizations which, instead of central-izing resources and investing in institution-building, commission individual research projects 'whenever they need something'.

Academics of this generation were often torn between political commitment to socialist ideals and professional frustration at being unable to produce and publi-cize the knowledge they felt could help to advance them. Some are thus perplexed by colleagues' defections from 'communist' to 'democratic' ideas, here meaning a new anti-Soviet rhetoric of democratization and neoliberal capitalism (Pine and Bridger 1997: 5). Asanova (2003) describes this as a 'great tragedy' and a betrayal. 'How did it happen', she wonders,

> that we so quickly 'forgot' about the decades-long preaching of communist ideology that we believed in as the 'sole truth' and 'sole science'? Is it proper that, not having clarified these painful and core questions for ourselves, we have begun to elaborate a 'new ideology' as if the former one did not exist, as if those people who now so energetically take up the ideology of 'national rebirth' or, let's say, the ideology of the 'all-consuming market' did not also militantly struggle for the realization of 'communist ideas'?
>
> (Asanova 1995)

Like others, she challenged the conflation of sociology and ideology during the Soviet period through a subjective interpretation of the spirit of that ideology. While she is critical of the repressive imposition of Soviet ideology and its dele-terious effects on intellectual creativity, she also maintains respect for the 'great idea of equality' that was fostered during the Soviet period. She, herself the author of a doctoral dissertation that was rejected as 'bourgeois ideology' in Kirgizia during the 1970s, is one of few academics to come forward with a critique of this transformation.

During the late 1960s and early 1970s, sociology was part of the Communist Party's broader 'hegemony of representation', described by Yurchak (1997) as 'a system in which all official institutions, discourses and practices are always-already produced and manipulated from the center as one unique discourse'. The experience of early social scientists in Kirgizstan, however, reveals that manifestations of this discourse varied widely, as did its effect on those who encountered it. For contemporary Kyrgyzstani sociologists who emphasize Tabaldiev's role in the establishment of sociology in the republic, this dissonance determined the fate of his laboratory and the discipline in the republic. This founding narrative has therefore taken on epochal qualities in both disciplinary histories and struggles for material and cultural resources, becoming metaphorical for academics negotiating a continuing tension between scientific truth and socio-political power.

4

KNOWLEDGE AND NATIONAL
IDENTITY DURING PERESTROIKA

Tabaldiev is not the only symbolically significant figure in the history of Kirgizstani social science, nor is the story of the KSU laboratory the only founding narrative of the field. In 1983, a second sociological laboratory appeared in the Frunze Polytechnic Institute (FPI), after independence renamed the Kyrgyz Technical University (KTU). Members of this laboratory consider its founder, Kusein Isaev, the true 'father of Kyrgyz sociology'. Indeed, Isaev has been more publicly vociferous in advocating for the field than Tabaldiev ever was, and his political campaigns to promote and defend the discipline during perestroika and following independence have earned him considerable professional status even among his ideological rivals. What differentiates the two men, however, is not essentially rooted in the types of work they produced or the methods with which they did so. Both were communists when they opened their laboratories, benefi-ciaries of the hierarchies of Soviet academic culture, and engaged in ongoing projects to carve out niches of intellectual freedom within a heteronomous system of knowledge production that they also continued to legitimate. Their difference as historical figures is rather a question of definition: what now constitutes 'real' and 'valid' social science in the society, and who has the authority to lay claim to its development?

Prior definitions of sociology in Soviet communism have repercussions for its contemporary post-Soviet legitimacy. For, within the dominant paradigm of scientific knowledge in the region, it can only be and have been objectively one thing: either a scientific method for discerning the truth about social reality, or a form of political ideology. Acknowledging the latter as part of the history of the field would contradict claims to its universal scientificity; hence, claimants for historical recognition must demonstrate the scientific credentials of their academic practice and conformity to or movement towards non-Marxist, functionalist norms of scientific knowledge production. The debate over the origins of 'true' sociology in Kyrgyzstan has therefore become highly political as members of the small circle of social scientists struggle to determine the boundaries of these definitions, and as a result, compete with one another for professional status. Those whose academic ancestry, as well as current teaching and research, fits into contemporary conceptions of legitimate science have greater intellectual authority and access to material and cultural resources than those whose genealogy is more ideologically 'suspect'.

This type of boundary-work is a common feature of the history of disciplines. The creation of space between socialist science and ideology had been an increasingly important element of the professionalization of sociology throughout the Soviet Union during perestroika. In the Central Asian republics, however, it also came to reflect and symbolize cultural tensions in Soviet identity on the periphery of the empire. The gradual association of social science with both nationalization and decolonization distinguishes the history of sociology at the FPI from its earlier predecessor, and within this.

Sociology during perestroika and the perestroika of sociology

Marxist sociology had become a visible, albeit perpetually beleaguered, part of the Soviet intellectual landscape by the beginning of the 1980s. Although not institutionalized like the more 'theoretical' subjects of philosophy and scientific communism (Zaslavskaia 1989: 111–13), a disciplinary journal was founded in 1974 (*Sotsiologicheskie issledovaniia*, or *Sociological Research*), sociology departments were organized in republican academies of science, and research centres employed thousands of people throughout the USSR (Zaslavskaia 1989). However, increasing intervention from the Communist Party, including a politically motivated restructuring of the Institute of Concrete Sociological Research in Moscow in 1971 and the silencing, censoring and deliberate under-utilization of research, meant that sociologists remained under pressure to conform to political imperatives while simultaneously demonstrating their scientific autonomy (Brym 1990; Weinberg 1992). These new pressures were coupled with a decline in actual support for research and a Ministry of Education ban on teaching sociology in higher education. During the 1970s, Soviet sociology thus entered into a state of suspended development.

This began to change during the early 1980s, when sociologists who had been displaced from their official academic posts began working as 'constructive dissidents', producing subtle critiques of both society and social science itself (Weinberg 1992: 4). The intellectual reform of sociology became discursively intertwined with the transformation of society, particularly as science was still considered the foundation of 'correct' development. There was a growing sense of urgency throughout the Soviet Union, including in Central Asia, about the need to recover the 'distorted' aims of socialism through a more critical interpretation of Marxist–Leninist principles to social work and intellectual analysis (Bekturganov 1990: 107; Ivanov 1988; Koichuev 1988: 5; Sherstobitov 1987: 5). Academics were called upon to eschew the dogmatism and redundancy which was said to have rendered knowledge production impotent during the post-Stalin years. In Kirgizia, they were instructed to use their critical faculties in the belief that this would, and in so far as this would, 'accelerate the socio-economic development of [the] country and further elevate the social sciences in Soviet Kirgizstan' (Sherstobitov 1987: 3).

Gorbachev's economic and social reforms reinforced this rhetorical revival of sociological purpose. Members of the intelligentsia and party leaders invoked excerpts from his political speeches in appeals to revive theoretical sociology, citing that 'theory is necessary... not only for perspective of social and political orientation [but] literally for every one of our steps forward' (Batygin and Deviatko 1994; Sherstobitov 1987: 3). In 1988, the Central Committee of the Communist Party issued a decree 'on increasing the role of Marxist–Leninist sociology in the resolution of the central problems of Soviet society'. This decree, seized upon by many as a catalyst for the development of sociology, called for 'the necessary strengthening of sociological work in branches of the national economy and in industries, increasing the role of social development services in the quest for productive labour reserves, the decrease of cadre instability, the administration of social processes in workers' collectives, and the planning of their social infrastructures' (Vlasova 1989: 41). In addition, new research themes began to emerge and sociologists supplemented conventional studies with inquiries into health and social exclusion, new forms of social cooperation, housing and health, crime and substance abuse, gender issues, social movements, ideology and socialization, youth and sub-cultures and postmodernity (Eades and Schwaller 1991; Weinberg 1994). Older models of social structure were tentatively challenged, as were critiques of 'bourgeois sociology' and Western Marxism (Batygin and Deviatko 1994).

Restructuring professional power and opportunities

Kirgizstani sociologists quickly pronounced commitment to the new political and economic reforms. However, they also saw in them an opportunity to establish sociology, as it had been defined in previous decades, in a new and more promising political context. The calls for greater efficiency provided ample opportunity for sociologists to represent their empirical work as significant and valuable for society, while the invitation to criticism made it possible for them to re-market the hitherto 'dangerous' elements of their profession as being perfectly suited to the new regime of *glasnost* [openness] in politics and science. For example, Tishin argued that

> [n]ow sociology, like no other social scientific discipline, can effectively and actually realize the positions formulated by M. S. Gorbachev at the All-Union meeting of chairs of departments of social science: 'science and theory are indispensable where and when the usual skills of action don't work, when past experience and practical native wit no longer give the needed advice, when principally new decisions and non-standard actions are necessary'.
>
> (1988: 62)

Earlier connections between science and modernization were reinterpreted in a new context: no longer as an aspiration to 'catch up' with the metropole but increasingly to articulate alternative forms of intellectual sovereignty within the Soviet system. Sociologists, however, continued lobbying for better access to education, training and resources in the more well-resourced centres of sociology in Moscow and Leningrad. By increasing the quality and usefulness of sociological research and teaching in Kirgizia, it was argued, the republic would hasten its transformation from a peripheral and 'backward' republic to a modern, autonomous and equal member of the Soviet empire. This in turn would subsequently raise the profile of Soviet sociology. The real or imagined possibility of a 'revolutionary renewal of Soviet society on the whole and in the union republics in particular' (Isaev 1991b: 32) raised hopes among Kirgiz and other Central Asian sociologists that they would finally be able to control the 'structure, work, concepts and financial organization' of their field (Isaev 1991b: 34). They believed this would enable them to be more influential in developing policies which were relevant to social life in Kirgizia, and which honoured 'national' traditions and ways of life while still providing a 'modern' scientific alternative to 'traditional' social relations.

In 1989, seemingly unaware of immanent upheavals in the social structure, sociologists in Kirgizia began to organize a Kirgiz branch of the Soviet Sociological Association (Isaev 2000; Isaev and Bekturganov 1990). They also began planning new-nationally based initiatives such as the Kyrgyz Union of Sociologists and a republican centre for sociological research which, while being affiliated with the Communist Party, was not subordinate to any central institution (Isaev and Bekturganov 1990). While they continued to work in industry throughout the 1980s, they also expanded their work in party organizations such as the *Komsomol* as the ideological offensive to promote sociology gained momentum. In 1983, the Communist Party of Kirgizia organized a centre for the study of public opinion, which was essentially designed to study the party for the party (Bekturganov 1990: 106). Several years later, sociologists from KSU cooperated with members of the Moscow State University's journalism department and national and republican publishers (*goskomizdat*) to conduct a major study of the regional press (Tishin 1989). In this and the following year, the USSR State Committee on People's Education passed a number of decrees granting universities the right to teach sociology and train sociologists (Isaev 1991b: 32).

At this time, the focus of rhetoric about social science shifted from technocratic administration to 'criticism' and 'self-criticism'.[1] The social sciences were among the first targets of critique. Sociologists declared themselves poorly qualified for the social role they were expected to play in reforming Soviet society. Despite assertions that sociology had assumed a new role for perestroika (Zaslavskaia 1989: 105), it remained highly empirical and associated almost entirely with service to state and party, as it had been since its construction as a Marxist science of society (Goldfarb 1990: 108). While the waning of repressive policies towards social research, the invitation for sociologists to provide administrative bodies with 'truthful' and 'accurate' data, the analysis of Communist Party resolutions from sociological perspectives, and the chance to offer 'feedback' to policy makers facilitated the establishment of greater autonomy, they were more immediately attempts to improve the way that social knowledge could be applied to further the interests of the power elite.

One crucial change occurred at this time in the relationship between sociologists and Communist Party authorities. Gradually, Kirgizstani sociologists began to define themselves as an alternative power base within the socialist project.[2] They no longer defined their role in 'assisting practice' (Zaslavskaia 1989: 117) as the mere 'scientific' confirmation of state or party decisions; serving the state no longer meant simply being subordinate to it. Instead, sociologists began to assert that they must play a proactive role in formulating political, economic and social policies and in analysing and criticizing those that proved to be ineffective. Sociology became redefined as a guarantor of glasnost and perestroika, an 'objective', 'scientific' and thoroughly Marxist antidote to the 'anti-socialist' abuses of power, which they argued had prevented sociologists from fulfilling their 'natural' role in assisting the planning, organization and management of the ideal socialist society.

In Kirgizia, philosophers and scientific communists who had advocated the development of sociology as an autonomous discipline since the early 1970s made this new position clear in public as well as in the academy, and in polemical articles on the subject in the popular party monthly *Kommunist Kirgizstana*. One, for example, stated,

> we believed that the socialists would succeed in building this project, and accepted the technology on the principle that the bureaucratic party apparatus has better knowledge of how the system must look. Not surprisingly, with this pragmatic approach and dogmatic conclusions, they began to interfere in social scientific research and the study of public opinion if they did not confirm earlier theory. It was precisely this approach that forced social science into scholastic theorizing and led to a crisis of the theory of scientific socialism.
>
> (Bekturganov 1990: 107)

Throughout the Soviet Union, academic elites launched a Marxist attack against the Stalinization of knowledge in previous years, challenging state hegemony by using the government's own rhetoric of free inquiry. The most prominent of these was Russian sociologist Tat'iana Zaslavskaia who, in 1986, addressed the Soviet Sociological Association with a scathing speech on 'the role of sociology in addressing the development of Soviet society'. She accused social scientists of 'bringing up the rear of society' in their obsequious confirmation of Communist Party ideology and challenged them to initiate rather than follow policy in the new era of economic and political restructuring (Zaslavskaia 1989: 105).

Similar critiques soon appeared in Kirgizia as well (Kakeev 1990; Tishin 1988). In 1987, the chief editor of the 'Social Science' series of the Academy of Science's journal published an article on the 'highest mission of the social sciences', urging a return to the spirit of Marxism to preserve the socialist project. The following year, the vice-president of the Academy of Science's Division of Social Sciences argued that 'the scientific base of perestroika is Marxism–Leninism' and that, because 'perestroika demands the creative alternation of the theoretical position of Marxism–Leninism through an analysis of modern social phenomena and ideological and economic decisions', social science was more relevant than ever (Koichuev 1988: 3). Promoting the value of scientific truth and struggling against its monopolization thus rose to the top of the agenda in professionalizing sociology (Tishin 1989: 4).

The perestroika of sociology in Kirgizia, however, also had more specificities, specifically those related to the 'national question'. One was the emergence of critiques of the republic's relationship with the Soviet centre, both in terms of its academic dependency and, more perhaps significantly, its intellectual colonization. Isaev (1998a) recalls that from the very beginning of perestroika, he anticipated the 'collapse of the united informational space and established methodological elaborations and literature, and struggled not to miss anything' that he could continue obtain from Moscow and Leningrad. In addition, while Kirgizstani sociologists reiterated criticisms of the historically 'unscientific' approach to policy making and governance across the Soviet Union and the general lack of intellectual freedom (Isaev and Bekturganov 1990), they also began drawing attention to regional differences in Soviet society and calling into question fundamental tenets of its organization (Isaev 1991b). For example, a report on the first conference of sociologists in Kirgizia, argued that

> [o]ne of the reasons administrative measures are not effective is the mechanical transferral of measures produced in other regions of the country to here. For example, in central Russia, particularly in the regions of Nechernozem'ia, where the rural population is ageing, the call for young people to remain in their villages is fully explicable. In our republic, on the contrary, there is overpopulation and unemployment in the villages, and young people have limited possibilities to choose a profession or activities. In such conditions, slogans that were until not long ago part of our official ideology – 'All graduates to the farm!' and 'Let the whole class stay on the kolkhoz!' were deeply mistaken.
>
> (Isaev and Niyazov 1990)

In Moscow, however, such differentiations were interpreted as divisive, and by the mid-1980s, high-ranking academicians (some within the republican structures) expressed concerns that social science in Central Asia not only suffered from 'all-union' afflictions such as the 'boring and dull repetition of truisms, fear of the new, and dogmatism', but that it also exhibited specifically 'national' problems such as 'a narrow mindedness of problematics, departing from regional and all-union significance' (Sherstobitov 1987: 4). The privileging of 'national' analysis over class analysis and the 'glorification' (or sometimes the mere mention) of

70

national historical figures were defined as evidence of this tendency. It was argued that 'at times, under the guise of national originality in a number of scientific works, efforts are made to present, in idyllic tones, reactionary-nationalistic and religious survivals, in contradiction with our ideology, socialist way of life, and scientific worldview' (Sherstobitov 1987: 4).

Some sociologists in Kirgizia, however, remained unconvinced that the uncritical use of universal Soviet categories was an effective solution to the problem of regional underdevelopment in the social sciences. Isaev, for example, began to develop his notion of a 'national Kyrgyz sociology' during this period, arguing that social science was more politicized in Kirgizia than in other parts of the Soviet Union because, he claimed, 'the distortion and deformation of the social–theoretical heritage was more pronounced here than in the center. . . . Marxist–Leninist social science, having not arisen on Kyrgyz soil, lost its critical edge and revolutionary nature under the strong pressure of Stalinist ideology and repression' (Isaev 1991b: 30).

The idea that a one-size-fits-all approach to the analysis and management of Soviet society had been detrimental to sociological understandings of life on the imperial periphery was heightened when, in 1990, disputes over the redistribution of property and position in the south of the country exploded into violent riots between ethnic Kyrgyz and Uzbeks. Sociologists argued that the repression of critical research into ethnic relations and the ideological mantra that there were no such tensions in socialist society had obscured the analysis, and hence the prevention, of ethnic conflict in Kirgizia (Nurova 2001), particularly as the party had censored studies which suggested that 'relationships between ethnic groups had been worsening for ten years before the 1990s' (Tishin 1998: 34). As perestroika progressed, the underdevelopment of sociology in Kirgizia became increasingly correlated with the underdevelopment of Kirgizstani society and the denial of national autonomy, identity and knowledge.

Sociology at the Frunze Polytechnic Institute

These new intellectual and political orientations emerged first from within a new sociological laboratory at the Frunze Polytechnic Institute (FPI). Isaev, then a prominent communist academic, established the laboratory in the Department of Scientific Communism, which he had founded and chaired in 1969 at the behest of Communist Party Secretary Togolokovich Murataliev (Abazov 1989; Group of Independent Sociologists 1993; Isaev 1998). In 1989, the laboratory was expanded into a Department of Sociology and Engineering Psychology in order to replace the Department of Scientific Communism, and in 1993 it was transferred to the Bishkek Humanitarian University (formerly the Institute of Languages and Humanitarian Sciences) to become part of the school's new Department of Administration and Sociology (later the Sociology Department), where it continues to operate (Isaev 1999a).

71

Unlike Tabaldiev, Isaev is renowned not as the founder of an academic institution, but for his role as a professional advocate for sociology and his efforts to create a national, specifically Kyrgyz, sociology. Like Tabaldiev, Isaev is a charismatic figure. During the Soviet period, he commanded significant authority both among his students and within the Communist Party, and has been called 'one of the greatest scholars' of his time in the republic (Sagynbaeva 2003). Many of his former students credit him with the single-handed development of sociology in Kyrgyzstan, particularly noting his role in nurturing a group of well-trained 'cadres' for whom he organized educational opportunities in some of the best academic institutions in the Soviet centre. One says,

> I don't always agree with [Isaev] on a number of methodological questions. But his role in the establishment [of sociology] is very important, because of all the candidates of science which we have today, 90 per cent are owing to him, during the Soviet period when it was only possible thanks to his authority. They were sent from Kirgizstan to study in Moscow, Leningrad, and Sverdlovst, to the very best schools of sociology. He did all of this. He went to the Ministry and made demands, I mean he stayed there and spent the night to demand.... He simply really wanted sociology to exist, so that there were specialists and so that these specialists received an education in good schools such as Moscow State University [MSU] and the Institute of Sociology within the Russian Academy of Science–he sent them there. All this was his personal work.
>
> (Anonymous)

This professional opportunity structure – a mixture of happenstance, curiosity and Isaev's intervention – was typical for young Kirgizstani sociologists of the time. Similarly, another recalls that

> [w]hen I became a sociologist long ago, it was by circumstance. After completing higher education...I could not find a job in my own specialization. They invited me...to the sociological laboratory at the FPI. Professor Isaev was the scientific director of that laboratory....And thus I went to work there; there wasn't anything anywhere else and, little by little, I started to learn more about this science. I worked there [some] years, earned my degree and defended my dissertation. I studied as an aspirant in sociology at the Russian Academy of Science in Moscow, then at the Academy of Science of the USSR, and thus I became a sociologist.
>
> (Anonymous)

Isaev, he says, had the greatest influence on him. He always said, 'come on, write; do some research, *tovarisch*...go here, go there; there's a conference, get an invitation to go'.

These efforts to recruit and train sociologists were relatively successful during the late Soviet period, not in small part because the field had gained a degree of legitimacy during perestroika. By the late 1980s, in fact, sociology was relatively well institutionalized in the USSR's major academic centres, and social scientists on the periphery were eager to benefit from and contribute to this trend. Although the FPI laboratory survived into the 1990s, however, Isaev's work contributed more to the popularization of sociology in Kirgizstan than it did to its institutionalization. This was partly due to the persistence of tensions between political and scientific responsibilities in sociological work, and to the continuing reluctance of political and economic elites to loosen their grip on their ideological control of social representation. In the late 1980s and early 1990s, Isaev and others tried again to reconcile these tensions, and the spectre of a more critical and analytical sociology appeared on the intellectual landscape. Nevertheless, this remained embedded within the colonial logic of state science, and many of its achievements were nullified abruptly with the collapse of the Soviet Union in 1991.

Disciplining sociology

One change that occurred during the 1980s and which was maintained after independence was the re-imagination of sociology from an extradisciplinary practice to a semi-autonomous or autonomous academic discipline. The FPI laboratory was initially organized as a research unit within the Department of Scientific Communism and defined as an 'instructional-auxiliary-sociological-laboratory' (Sydykova 1998). As such, it fulfilled similar functions to the KSU laboratory by integrating education, professional training and political service. Isaev used it as a base for training the 'second generation' of Kirgizstani sociologists, who, unlike those who joined the KSU laboratory in the late 1960s and early 1970s, entered a nascent academic discipline as opposed to an extradisciplinary practice.

Frustrated with the republic's dependence on the centre, Isaev supported new Soviet policies to increase the production of 'national cadres' in the Kirgiz academy. The laboratory provided a site for his long-term project to create a critical mass of professionally trained, self-reproducing sociologists who would be qualified to conduct empirical research, teach sociology in universities and contribute to the field's overall institutionalization and professionalization. He also placed new emphasis on distinguishing sociology as an independent discipline, distinct from but compatible with scientific communism and historical materialism. According to Sagynbaeva (2003), the laboratory's first home within scientific communism was problematic for Isaev as he wanted to produce 'specialists who would actually be pure sociologists'. The notion of disciplinary 'purity' has led many sociologists to consider this laboratory and not Tabaldiev's to be the 'first' sociological institution in the republic (Ibraeva 2003; Osmonalieva 1995). This is particularly true of Isaev's former students, the first generation of students that could systematically defend dissertations in sociology (albeit not in Kirgizstan), pursue careers as academic sociologists and take advantage of resources in newly established departments of sociology in Russia. In fact, according to Isaev (2000), the institutionalization of sociology in Kirgizia only 'beg[an] with the preparation of professional specialists in the scientific centres of Moscow, St. Petersburg, and other cities in the RSFSR, in which more than fifteen candidates of sociological science were trained in the 1980s and 1990s'. Ironically, however, the price of increased recognition was also increased dependence. The rhetorical redefinition of sociology as an independent field did not significantly alter the political roles which had been ascribed to it in preceding decades. The laboratory's organization and thematic foci remained dictated by economic and political forces in Soviet society.

Isaev's position in the party, belief that sociology should support the improvement of socialist planning and relatively non-threatening research interests (such as rural–urban migration) meant that there was little tension between him and the administration over control of the laboratory (Isaev 1993a). In fact, his ability to combine new discourses of democratization with official party rhetoric on economic and social development secured the laboratory's survival during the mid-1980s amidst growing fears elsewhere that the discipline harboured 'subversive' tendencies.

Sociology and social planning

During this period, Kirgizstani sociologists also capitalized on Moscow-led initiatives that encouraged the use of 'complex research' in social planning. Discourses on socialist development shifted from economic productivity to a more holistic conception of reform, which defined social and economic development as mutually enhancing. The new theoretical focus on the significance of 'the social' (ways of life, traditions and particularly 'public opinion') created spaces for sociologists to strengthen their positions in industrial institutions and create new roles for themselves beyond the factory and farm.

The ideological products of the 27th session of the Central Committee of the Communist Party in 1986, as well as Gorbachev's speeches on the relationship between social science, social planning and development, figured heavily in shaping the development of policy research in Kirgizia during this period (Tishin 1988; Tishin et al. 1989; Vlasova 1989). While sociologists were eager to be included in policy making, the weakening of censorship also enabled them to challenge ideological forms of Soviet policy and advocate an even greater role for themselves in defining the meaning of perestroika.

Throughout this period, the FPI laboratory remained firmly integrated into the administrative apparatus of state and party, with many of its research projects conducted specifically on zakaz (commission) for governmental organizations seeking data to inform social planning. Like the KSU sociologists who had previously worked in cooperation with industrial managers, members of Isaev's research team were oriented towards gathering information obtained through quantitative research and making 'scientifically based' recommendations for managerial and disciplinary action (e.g. Dzhangirov *et al.* 1987). However, as perestroika progressed, their research became increasingly nationally oriented and included features unknown in previous decades, such as discussions of indigenous social problems created in part by inequalities within Soviet society. It also revealed subtle changes in the status of sociological research during this period, in particular a convergence of sociologists and the republican power elite. The mere fact that sociologists produced and distributed empirical evidence of social problems signalled a loosening of political controls on the discussion of 'negative' phenomena and a growing willingness to at least formally consider 'public opinion' and subjective experience as sociological 'data' (Dzhangirov *et al.* 1987). It also anticipated new connections between social science, public opinion and ideals of democratization, which were at the time only beginning to emerge (Abazov 1989; Toktosunova and Sukhanova 1990). As Isaev and Bekturganov pointed out,

> in the years of repression, the bureaucratic apparatus of government laid down its veto on the study of all negative social phenomena and processes, the revelation of which could expose it in the people's eyes. They carefully concealed the negative aspects and intentionally circum-vented acute problems of social policy, international relations, inde-pendent religiosity and etc., which demanded a principled and critical evaluation.
>
> (1990: 3)

It was not until several years later that such critiques would spawn new discourses linking social social scientific knowledge with political independence, democracy and truth telling, thereby altering the definition and role of sociology in Kirgizia.

From the 'national question' to 'national sociology'

In 1989, the Department of Scientific Communism and sociological laboratory were combined into a new department of Sociology and Engineering Psychology (Isaev 1993; Osmonalieva 1995). This was a significant change in sociology's position within the system of academic disciplines, as it was ascribed semi-autonomous status in relation to other disciplines such as engineering psychology (concerned with the social and psychological aspects of scientific management and social planning). As a prominent member of the Communist Party, Isaev remained loyal to the goals of rational social planning; but invoked the vocabu-laries of glasnost and perestroika to criticize the Russo-centric bias of many poli-cies applied to Kirgizstani society and capitalized on weakening intellectual controls to publish articles about social problems in rural communities. The FPI laboratory was one of the first institutions in the republic to make perestroika itself into an object of sociological analysis and advocate that 'the necessity of including results from sociological research in social administration requires that

[we] develop the problem of activating the human factor [in order to] realize the principles of social justice and the consolidation of socialist ways of life' (Isaev and Bekturganov 1990: 7). In the late 1980s and early 1990s, members of the laboratory began to conduct opinion polls about privatization, local governance and Communist Party reforms, and elaborated basic (and contested) methodologies for 'rating' politicians standing in local and republican elections.

Other sociologists used a similar rhetorical strategy, linking the development of sociology to socialist democratization and market liberalization. In 1990, for example, then-director of the party's Centre for the Study of Public Opinion argued that

> due to the disengagement of political power from public opinion and real life processes, the break between the political-economic structure and social expectations has not only not decreased, but continues to increase. We can only find a way if we concretely and, at the same time, complexly study and analyze the real complex situation by applying Marxist methodology. Only then can we make political and state administrative decisions that are oriented toward a democratic society, deepen the transformative process in civic and political life and realistically measure the forms and methods of administration in society.

> Quality and in-depth public opinion research would allow a more accurate and clear definition of the priorities and ideals of a reformed, human and democratic socialism.

> (Bekturganov 1990: 107–08)

Despite these subtle changes to definitions of socialism, Kirgizstani sociologists aspired to be more rather than less Soviet during perestroika. A survey conducted among those attending the first conference of sociologists in 1990 suggested that the majority were most interested in three major sub-fields within sociology: the sociology of 'nations', economic sociology and the sociology of labour, and the sociology of youth (Isaev and Niyazov 1990: 150). And yet, there emerged a double-edged criticism of sociologists at this conference. On the one hand, it was argued that social scientific knowledge in Kirgizia was not sufficiently 'national' in focus; it was too deferent to Russia, too 'abstract' in its generalization. On the other hand, some asserted that national traditions and 'backward thinking' prevented social scientists from liberating themselves from the habit of reproducing dogmatic Marxist–Leninist platitudes. There was also considerable criticism of the 'rudimentary' institutional and intellectual state of the discipline, which, it was argued, had made little progress since the establishment of Tabaldiev's laboratory in 1966 (Blum 1990; Isaev and Niyazov 1990). The question of intellectual colonialism and academic dependency had finally been brought to the fore of the professional community – thought they remained couched within the logic and language of colonialism itself.

Finally, it emerged that Kirgizia was the only republic still lacking a separate branch of the Soviet Sociological Association. Isaev's deliberations to create one began at this conference, where he was elected president of the short-lived endeavour. The elaboration of a national or 'Kyrgyz' sociology did not coalesce until after independence. However, the genealogy of this focus on the republic as a geopolitical unit of analysis is evident in many of the publications produced by the FPI laboratory in the years immediately preceding independence. The 'nation', now meaning the republic and not the USSR, soon became a central feature of his own theoretical and empirical work and shaped the direction of research within the laboratory. In addition, his reputation as a social critic and member of the political opposition (the latter a label which he rejects, preferring to call himself a 'patriot–opponent') can also be traced to this period (Isaev 1998). Even more than Tabaldiev, Isaev (1989) has been critical of the unequal relationship between the peripheral republics and the Soviet centre and of the ideology of *sblizhenie*, arguing that cultural development in the periphery was stunted specifically because it 'excluded all concepts of national development and national pride'.

Instead of blaming this entirely on Russian dominance in the region, he criticized passivity within Kirgiz culture. He drew on the work of a renowned Kirgiz writer to compare republican intellectuals to the fictitious *mankurts*, semi-literate and incompetent prisoners of war who became mindless slaves after having their heads bound in camel skins (see Aitmatov 1983). In arguments recalling those of anti-colonial critic Frantz Fanon (1963), he also claimed that the 'national intelligentsia are drawn from the peasantry and quickly move to become bureaucrats, directed by the center … they are good at mimicry and have rejected all things national to please the center' (Isaev 1989). However, Isaev advocated the development of 'national self-consciousness' not as a political challenge to the Soviet state, but as a method for equalizing political relationships within the multi-national empire. As he argued in an article published just before independence, and in some sense echoing earlier exhortations of Tabaldiev and his colleagues,

> particularly in the Central Asian republics, where economic backwardness combines with cultural particularities, bloody conflict has broken out. And it is here that sociology can and must render an invaluable favor, for it can prevent a society from possible social tension, give concrete recommendations and determine the path of their resolution.
>
> (Isaev 1991b: 27)

In the years following independence, Isaev's on the relevance of sociological knowledge for social problems arising from strained colonial relations rapidly evolved into a new discourse on a specifically 'national' sociology. However, this was concurrent with increasing interest in the 'universality', 'neutrality' and 'objectivity' of a more scientific sociology, whose development was deemed essential for raising academics' professional status both in Russian science and on the periphery. Boundary-work then focused on creating theories of knowledge that enabled the production of universal scientific truths for particular social circumstances – an epistemological complexity that has never yet been fully articulated.

The challenge to generalized Soviet sociology in Kirgizia

The 'bloody conflict' mentioned earlier referred to the week-long riot which erupted between ethnic Kyrgyz and ethnic Uzbeks living in the south of the republic in June 1990.[3] The 'Osh conflict' was condemned by Soviet authorities as a 'terrible misfortune' and the result of young people 'giving way to their emotions and [being] stirred up by the ambitions of extremist-minded elements' (Appeal Central 1990) among a 'people who, for centuries, [had] lived together in peace and harmony' (Appeal USSR 1990). Kirgizstani sociologists, however, interpreted the incident as a glaring indictment of structural injustice within the society and a consequence of years of denying the existence of ethnic tensions in the republic. It was also seen as part of a larger trend of violent demonstrations against political repression and economic dissatisfaction throughout the USSR (e.g. in Kazakhstan, the Baltics and the Caucasus), and one sociologist argued that 'if ideological work in the sphere of national relations goes on without deep scientific analysis of the real situation of national processes, without an account of the opinion and mood of the representatives of various nationalities and peoples, then it will lead to the appearance of national egoism and arrogance, to national isolation and particularity [and] to dependent moods and parochialism' (Bekturganov 1991).

Ultimately, the events raised awareness that Kirgiz society was harbouring unresolved problems, many of which could not be attributed to 'regularities' of Marxist theories of development in the all-union context or dealt with within the conventional theoretical formulae of Marxist–Leninist philosophy. Cracks had begun to show in ideological pronouncements about the stability and high quality of life on the periphery. As the pace of social change quickened at the Soviet centre, intellectual life in Central Asia became more dynamic, with more public discussion of social problems accompanied by greater attempts from political leaders to deny them. Sociologists called greater attention to these issues in the media and increased demands that they be studied empirically.

While Kirgizstani academics remained heavily dependent on Russian academic institutions and continued to work within the Marxist frameworks which constituted the bulk of their theoretical knowledge, they aspired less to orient this knowledge outward toward the abstract problems of a generalized 'Soviet' society. Isaev in particular turned a critical gaze on Kirgizstani society and, as a result, on the colonial logic of Soviet social science, which he argued had prevented social scientists from genuinely understanding their own society.

By 1991, sociologists had redefined their position within the Soviet scientific community and *vis-à-vis* the centre. They argued that by attaining relative autonomy to engage in research about problems of republican as opposed to generalized 'Soviet' concerns they would be able to make more relevant contributions to an increasingly pluralistic Soviet sociology as a whole. This in turn would allow them to muster greater support from state and party, which they argued would consequently stimulate the theoretical and methodological development of the field and hasten its professionalization (Isaev and Bekturganov 1990: 3).

The waning of generalized Soviet sociology in the Kirgizstan and the emergence of a more nationally oriented discipline, however, was not initially associated with aspirations to autonomy from political institutions or secession from the Russian centre. Republican sociology would suffer from the decentralization of Soviet science. On the eve of independence in 1991, there were no fully independent sociological institutions in Kirgizia. Academics remained dependent on subsidies from the central government and the inconsistent flow of commissions for research from the Communist Party and sectors of the 'national economy' (e.g. factories and state and collective farms). While books and pamphlets on Marxist philosophy abounded in university and public libraries, there was a paucity of literature on sociology; this was generally obtained through individuals attending conferences in the RSFSR and other more 'Western' Soviet republics (or, towards the end of perestroika, even abroad). In addition, even basic information on new developments in the discipline was only available in the capital city of Frunze, not in more rural regions. While the Ministry of Education had stipulated that sociology should become a required subject for university students in the early 1990s (Isaev 1998b,c), only a handful of individuals trained in Russia and the Ukraine were qualified to teach undergraduate sociology and universities were slow to implement courses. In early 1991, therefore, the recognition and support of national sociological communities within the broader framework of a reformed Soviet sociology held enormous promise for many social scientists in Kirgizstan. The immanent collapse of the Soviet Union did not.

Sociology's second-generation into independence

Unlike many academic institutions, the FPI Department of Sociology and Engineering Psychology survived late socialism. Shortly after national independence, Isaev was invited to a professorship at the Bishkek Humanitarian University and in 1993 the FPI sociological laboratory relocated to become part of a new department of Administration and Sociology at this university. It soon became recognized as the new base for professional training in the republic. Isaev continued to conduct research about conventional topics such as the adaptation of rural Kyrgyz youth to urban life and increased the frequency of opinion polls on privatization and political ratings, which became more critical as the country's politico-economic situation deteriorated. He says of this period, 'outwardly, all seemed fine – who would have expected that today we [would] not have the means to exist?' (Sydykova 1998)

On the one hand, Isaev's persistent efforts to institutionalize sociology throughout the Soviet and post-Soviet periods has made him an easy target for criticisms of opportunism and hypocrisy, particularly from sociologists formerly associated with the KSU laboratory. Tishin *et al.* (1998), for example, question the authenticity of his academic qualifications by pointing out that that he 'was head of a department of scientific communism for twenty-four years...and only after the collapse of communism did he become the "father" of sociology'. On the other hand, many others, particularly those associated with the FPI laboratory, interpret changes of intellectual and political position as a positive good. From this perspective, he serves as a role model for the development of more dynamic and pluralistic approaches to sociological thought. 'Kusein Isaevich really worked on the development of sociology in Kyrgyzstan', says one of his colleagues. 'He studies all the contemporary concepts and paradigms of sociological theory in France, England, Germany, and America. He thinks that we need to create our own sociological theory'. While few others actively support his project to establish an indigenous form of sociology (as one put it, 'maybe a national *association* or some kind of group of sociologists... but a national *sociology* – I don't think so'), his work provides them with a new, non-Soviet and 'national' point of reference with which to understand their own historical identity – an emerging identity which was abruptly ruptured in 1991 with Kyrgyzstan's declaration of independence from the Soviet Union.

5

SOCIAL SCIENCE AFTER COMMUNISM

In August 1991, the Kirgiz Soviet Socialist Republic declared independence as the sovereign Republic of Kyrgyzstan. Independence was widely unanticipated, but publicly celebrated with as much enthusiasm as if it had been won through popular struggle. Two months after Gorbachev resigned, the new Kyrgyz president, previously a staunch supporter of the liberalizing Soviet state, addressed the United Nations General Assembly, saying, 'now that the center has collapsed under the weight of the crimes it committed against its own people, there is no holding back the will of the republics which have found their freedom in a bid for political and economic independence' (Akaev 1991; see also Spector 2004). Sociologists clambered onto the bandwagon, believing then as they did later that 'in the conditions of an independent Kyrgyzstan... the possibility for the gradual development of a national sociology appeared' (Isaev 1998b).

Postcolonial, neocolonial, independent?

Independence did indeed alter the trajectory of the field's development, though not as expected. The disintegration of the Soviet Union transformed the intellectual, cultural and political context of social science in the communist bloc and necessitated a massive overhaul of the structure and organization of the disciplines in each of the constituent republics (Batygin and Deviatko 1994; Eades and Schwaller 1991; Kürti 1996; Ruble 1993; Skvortsov 1993; Tishkov 1998; Weinberg 1992, 1994; Zaslavskaia 1989). Social scientists contended with problems affecting academe throughout former Soviet space: institutional poverty, a lack of qualified instructors and teaching materials (including things such as paper for publishing books and journals; see Naby 1993); brain-drain and sale of grades (Aldasheva 2003; Asenbekov 2003; de Young 2001; Karim Kuzu 2003; Obychnyi prepodavatel' 2000; Osorov 2002; Phipps and Wolanin 2001; Reeves 2003; Tishin 1998). The experiences of sociologists in Kyrgyzstan clearly reflect six trends that Sabloff has argued characterize post-Soviet higher education as a whole, including increased access to education and decreased support for it, widespread pressure to 'Westernize' curricula and teaching methods, the erosion of faculty salaries and brain-drain, demoralization and exhaustion among educators, changes from specialized to 'flexible' curricula and the need to find new, non-socialized sources of income such as business, tuition fees and international organizations (Sabloff 1999: xi–xvii; see also Bronson *et al.* 1999).[1] These are compounded by less visible challenges for social scientists, such as personal poverty, loss of status and prestige, breakdowns in professional relationships, existential and intellectual crises, and broken or interrupted careers – though experience of each of these varies widely within the community.[2]

In Kyrgyzstan, independence initially created a sense of increased intellectual freedom. The epistemological orthodoxies of Marxist philosophy were dismantled as sociologists began to engage more openly with other schools of thought. Isaev (1991), for example, recalled that

'trips across the border, the study of works of foreign authors and of [his] compatriots, and books that were previously inaccessible' led him 'to conclude that Marxism is only one branch of social thought. There is a wealth of other views in the world'. He instructed his colleagues: 'open up for yourself once again Kautskii, Trotsky, Bukharin, Rykor, Chinov, and you will understand how poor and one-sided our own vision was.... Unfortunately, we not only had a false consciousness, but in principle an unscientific one'.

Abruptly, 'pluralism' in both theory and politics was, at least rhetorically, elevated to a new level of virtue (Isaev 1999a: 8, 2000; Kydralieva 1998: 171). Sociologists were ostensibly free to develop new theories of class, ethnicity, culture, stratification, power and social change in Kyrgyzstani society. However, the long dominance of Marxism–Leninism and the paucity of alternative theoretical scholarship prevented this freedom from intellectual monism from translating into a freedom for something else. As later pointed out, many felt that 'when the Soviet Union collapsed and the sole scientific knowledge of Marxist–Leninist history made the sociological approach seem useless, the social sciences began to suffer from uncertainty' (Isaev 1999e). When the intellectual architecture was categorically delegitimized by its very association with the Soviet past and declared incompatible with anti-Marxist and anti-socialist theories of society and social change. Sociologists were thus left with few conceptual resources. Bekturganov *et al.* (1994) have in fact argued that 'Kyrgyz sociology does not yet have its own requisite theoretical–methodological equipment that corresponds to local conditions' and that 'therefore, no one can intelligently explain the processes going on in the country'.

Independence therefore did not usher in a new era of intellectual confidence in Kyrgyzstan, and while the Soviet project was abandoned, many academics continue to defer to Russia and, increasingly, 'the West'. With limited access to new academic work and rising demands to gain fluency in these very ideas, many sociologists turned to classical writings in *zapadnaia sotsiologiia* [Western sociology] for guidance. Textbooks in Western classical sociological theory (otherwise known as the 'history of sociology') had been circulating since the 1980s, were published widely in Russia and were increasingly imported by foreign lecturers in Central Asia. This confirms, as Nesvetailov (1995: 61) points out, that within postcolonial science 'the major specific trait of the periphery is its dependence on the centre. This position has been retained by the former republics of the USSR. The only change has been the center's address: instead of the Soviet structures in Moscow, the address has become the world centers of scientific activity'.

Attraction to 'world sociology', as this is called in Kyrgyzstan, was rapidly embraced as the optimal model for the development of sociology as an academic discipline (Baibosunov 1998; Isaev 1993). This is visible in the types of theory that have become ascendant since independence and in the desire to 'internationalize' indigenous sociology so that it meets 'world standards' as well as local concerns. According to Tishin (1998: 34),

> the Marxist–Leninist theory of nations and national relations was refined with new worldviews by L. Gumilev, V. Mezhuev and V. Tishkov. The ideas of Max Weber, E. Viatra, K. Nurbekov and others received wide circulation. . . . Sociologists in Kyrgyzstan paid special attention to the views of English researcher E. Gellner and American sociologist S. Huntington. Ethnic processes, national conflict and other phenomena of a multinational society were studied anew by sociologists E. Elebaeva and N. Omuraliev, and ethnographer A. Asankanov.

Functionalists such as Parsons, Merton and Smelser have been extremely popular among sociologists (Isaev and Abylgazieva 1994; Isaev, Akmatova and Sharshembieva 1996), and Bourdieu and Habermas receive increasing attention (Baibosunov 1998). Specializing in a new field of knowledge, *konfliktologiia*, integrates American conflict resolution with Soviet theories of ethnicity. Giddens' 'third way' theories are combined with the more culturally familiar convergence theories of Sorokin, Aron and Bell (Isaev 1993c); Huntington's theory of the 'clash of civilizations' has merged with Sorokin's theory of cyclical history (1993c) to produce 'Eastern' or 'Asian' theories of development; and Beck's 'risk society' is used to theorize and criticize the human consequences of Soviet environmental policies in Kyrgyzstan (Isaev 2000). Isaev (1997) has also drawn comparisons between Popper's and Gandhi's theories of the 'open society', challenging the predominance of its 'Western' formations and recommending greater attention to more 'Eastern' ones. Externally produced social theories are not adopted undiscerningly, however, and their meaning is often transformed as they are integrated into existing epistemological frameworks. For example, structural-functionalist theories have been applied to traditional Marxist–Leninist themes such as class relations, marriage and the family, and national relations (see Isaev 2000; Isaev and Abylgazieva 1994; Isaev, Akmatova and Sharshembieva 1996; Shaidullaeva 1992).

As theoretical diversification is explored within severe material constraints, it is often represented as part of an intellectual crisis compounded by the institutional structure of science. Macwilliams (2001) dubbed the 1990s a 'decade of more freedom and less money' for Russian universities (see also Graham 1998); in Central Asia, the Soviet collapse led not only to the disintegration of Kyrgyz sociology's emerging structure and raison d'être, but also to 'the collapse of traditional links with great Russian educational institutions, limited access to Russian literature on sociology, the stagnant isolation and decline of standards and quality of diplomas for candidate and doctoral degrees' (Ibraeva 2003). Research centres in schools and factories dissolved and their personnel were scattered throughout the republic. Many prominent and accomplished ethnic Russian scholars left the republics; many others abandoned the academy in search of livelihoods elsewhere (Egorov 2002). Work became increasingly constrained for those that remained after Soviet subsidies for research and education were withdrawn; the Kyrgyz government has given only token support for social scientific activities since independence.[3]

Lack of financial support, human resources and teaching materials were also pressing concerns during perestroika (Bekturganov 1990; Bekturganov and Isaev 1991; Blum 1993; Isaev 1991b; Isaev and Niyazov 1990; Tabyshalieva 1986; Tishin 1980; Vlasova 1989). However, although they struggled to secure funding during the Soviet period and could do so only for applied or 'practical' research as opposed to theoretical studies (Isaev 1991a; Zhivogliadov 1990), many social scientists received regular commissions from industry, the Communist Party, and the state (Bekturganov 1990: 110). In 1988, for example, the Institutes of Economics and Philosophy and Law under the Kirgiz Academy of Science 'switched to working on *goszakazy* [state commissions] for the government...on a variety of problems that have important national significance', including drug addiction and homogenizing *vospitanie*; this was deemed 'the most suitable forms of linking science to production' (Koichuev 1988: 8). Similarly, in 1990, plans to create a republican centre for sociological research were based on the 'principle of serving the *zakazchik-ispolnitel'* [commissioner–user]', specifically the state and Communist Party (Isaev and Bekturganov 1990: 8).

Some argued that this strategy for funding social science – a peculiarly late-socialist combination of state regulation and market competition – further restricted intellectual and scientific autonomy. The trend towards *goszakazy* and increased demand was, after all, also a movement towards greater dependence on central power. At the turn of the decade, one Central Asian journalist wrote, 'given that there are no [autonomous sociological] institutions in the country', academics had two choices: either to offer their services to industry 'all in the hope for a crust of bread', or, 'for those who value independent thought and freedom of scientific enquiry, to set up [their] own cooperative and fill orders from industries, organizations and institutions on contract' (Blum 1990). Nevertheless, just prior to independence, the *goszakaz* system was for the most part viewed as a progressive development in the institutionalization of Soviet sociology in Kirgizia.

The break with these constituencies and the demise of the Communist Party as a political and economic force in Kyrgyzstan therefore translated directly into the total loss of sociology's funding base and clientele. Central Asian sociologists cite lack of support 'from anywhere' for teaching, research, publication or travel as obstacles to institutionalizing the discipline in the post-Soviet period (Asanbekov 2003; Blum 1990, 1993; Isaev *et al.* 1993a); lecturers' salaries now range from $15 to $250 per month.[4] However, since the late 1980s, foreign governments and international and non-governmental organizations have invested 'many millions of dollars' in reforming social scientific research in the former Soviet republics (Ruble 1999). The sponsorship of organizations such as the Open Society Institute, USAID, The Eurasia Foundation, UNESCO, Save the Children, the International Monetary Fund, the World Bank, the MacArthur Foundation, the International Labor Organization and others has enabled Kyrgyzstani sociologists to conduct research in otherwise impossible conditions (Ablezova 2003; Bitkovskaia 1996; Blum 1993; Ibraeva 2003; Isaev 1993a; No borders 1999; Osmonalieva 1995; Sagynbaeva 2000). Such organizations have in fact become the primary sources of funding for sociological research in Kyrgyzstan today.

The integration of sociological research into these new 'development' institutions and its increasing marketization is seen as both a blessing (No borders 1999) and a curse (Baibosunov 1993; Isaev 1996a; Tishin 2003). Dependence on these new *zakazchiki* also has an underside, and intellectual autonomy is still constrained and regulated by the new clientele. Post-independence sociology in Kyrgyzstan has also become highly commercialized and commodified. While the organizations often promote values of democracy and 'open' or 'civil' society, and capitalism by implication, their relationship with academics has at times been anything but egalitarian. Not only topics of research, but also research questions and design may be prescribed. In Kyrgyzstan, for example, sociologists are hired to gather data through surveys and interviews which are then analysed and published, often in English and in the form of institutional reports as opposed to scholarly papers, outside the country (Asanbekov 2003). Indigenous scholars have few rights to use this data for their own research purposes (Ablezova 2003; Nurova 2003; Omurkulova 2003). Independence from the Soviet Union, for example, entailed the recolonization as well as decolonization of social science, and a shift from intellectual colonialism to academic dependency and post- and neocolonial forms of knowledge production.

Furthermore, because there is little protection of intellectual property rights, there is limited knowledge sharing within the academic community. Some attribute this to secretive hoarding habits acquired in 'Soviet times' (Ablezova 2003), but there is also a more immediate and pragmatic concern that competing researchers or institutions will 'steal' research designs and instruments and thereby gain a competitive edge in grant proposals and commissioned projects from international organizations. When the country director of a major US governmental scientific organization attempted to set up a national database for sociological questionnaires, study results and data sets in 2001, for example, she met with great resistance from colleagues who, she felt, were 'very territorial about data and only wanted to sell it' (Omurkulova 2003). The Kyrgyz government, for its part, did not want to establish a national database within a dominant foreign organization.

Some veterans of the *goszakaz* system have even deeper reservations about the potentially subversive politics of institutional alliances between national sociological research and foreign development organizations, including espionage. Isaev (1998), for example, proposed that

> [w]e are in a rather interesting situation. For the past five to six years, foreign foundations have been financing many of the research projects by our local group of sociologists, aimed at gathering data on public opinion. They never publish their research results. Meanwhile, they have managed to gather strategic information, which our government and state institutions are unaware of. This implies that other countries have learned about our country's strengths and weaknesses, our market and economic potential, and how we think and what we think about. The ultimate threat, I believe, is in this phenomenon.
>
> (see also Isaev 2003; Sydykova 1998)[5]

Most academics, however, are more immediately concerned with its consequences for their everyday practices, including pedagogy. Although the formal ban on teaching sociology in Soviet universities was lifted in 1988, the Ministry of Education did little before 1991 to incorporate sociology into the university curriculum. After this time, following Moscow, it felt compelled to do so, and lost control over new private institutions that were created outside of the traditional educational frameworks. After independence, therefore, numerous departments of sociology sprang up in colleges and universities in the republic's capital. Some are given meagre support by the state, while others are sponsored by foreign governments and private donors. For the most part, however, there is low investment in institution building projects and many donors prefer to sponsor professional 'training' programmes for individual sociologists. This has left many scholars, particularly those untrained in the English language, with inadequate means of support for university work. Universities have thus become spaces of competition, with each promoting a new model of sociological education in an attempt to attract students and resources.

The proliferation of departments, programmes and research centres has also led to concerns that poorly qualified sociology instructors, many of whom are trained in Marxism–Leninism and scientific communism) produce unqualified graduates, who in turn assume positions in 'amateur' research companies, compete for contracts with international organizations, and deliver misguided information about society for public consumption. Soviet-era concerns about the detrimental effects of 'amateur' sociology on the discipline's professional status have been exacerbated by the decentralization of training, standards and resources in the independence period (Isaev 1998c); from complete state domination to professional fragmentation. There is still no body qualified to grant doctoral degrees in sociology in Kyrgyzstan, and those candidates who can afford to travel are forced to defend their dissertations in nearby Kazakhstan or in Moscow (Baibosunov 1998). Although some instructors have furthered their education through self-study, local pedagogical conferences and foreign exchange, most feel that they are often asked to teach well beyond their actual competencies.

Early enthusiasm about the promises of independence for sociology in Kyrgyzstan has thus been tempered by concern about the deleterious consequences of the decentralization and commercialization of science and education. The decades-long project to institutionalize Soviet sociology as a relevant and legitimate form of knowledge, in the service of both scientific truth and political commitment, was interrupted in 1991. Political independence did not automatically translate into intellectual autonomy, but rather into new forms of cultural and structural dependency that must be negotiated within an entirely new set of conceptual frameworks and institutional arrangements. Despite these constraints, however, belief in the possibility of modern scientific knowledge persists, and there is an almost unilateral consensus that social scientists can and must contribute to the 'transition' of society from 'totalitarianism and communism' to 'democracy and capitalism'. They should, according to its advocates, 'help advance the goals set by the government and president for the creation of a free, democratic and civilized society' (Isaev 1991a), provide 'accurate information' in order to stem the flow of destabilizing 'rumours' in society (Migration 1992), serve as a 'believable source of social information for making decisions or correcting the political behaviour of leaders' (Isaev et al. 1994b), 'strengthen the scientific basis of politics' (Isaev 1995), 'facilitate the skilful administration and development of society on the whole' (Isaev 2003) and 'analyze and differentiate contemporary politics, not leaving the sphere of political production solely to individual politicians, and in order to escape from the symbolic, or even outright manipulative thrust of certain points of view' (Isaev et al. 1997a).

Thus, as during the Soviet period, a tension between 'truth' and politics preoccupies efforts to institutionalize and professionalize sociology. Institutional, intellectual and political legacies of Soviet social science intersect with post-Soviet discourses of both science and the nation to shape the contours of this negotiation. Factors such as a collective disavowal of sociology's technocratic and ideological identity, contemporary demands for social scientific knowledge to be immediately 'practical' and 'relevant', scepticism about the effects of illegitimate power on knowledge, faith in the possibility and promises of scientific method, and competition for access to the highly competitive and unstable pool of professional resources available for academic work have intersected to create a new context for the production of discourses about the emergence, development and future of sociology in Kyrgyzstan. In this context, the question of whether independence necessitates a new 'national' sociology is high on the agenda.[6]

National sociology in Kyrgyzstan: myth or reality?

The creation of a national Kyrgyz sociology, or at least a sociology that is responsive to the needs of Kyrgyz society, is often portrayed as a uniquely Kyrgyz project. However, nationally situated sociological traditions (e.g. French, German, British, American) have been institutionalized in the history of ideas, deliberately national sociological communities were formed in decolonization movements (Akiwowo 1999; Ganon 1965; Hiller 1979), and the very notion of 'national' or 'indigenous' knowledge has sparked fierce debate among philosophers, sociologists of knowledge and development institutions (Agrawal 1995). Within postsocialist space, many other countries also initiated such projects after the Soviet collapse – in Eastern Europe (Keen and Mucha 1994), Kazakhstan (Luk'ianova 1990), Russia (Batygin and Deviatko 1994), Uzbekistsan (Toschenko 1998) and elsewhere.

In addition, sociology was historically a 'national' science in the Soviet context – but the definition of 'nation' has changed dramatically. The concept of *otechestvennaia nauka* [national or patriotic science] communicated both geographical and political meanings. While this may seem incompatible with contemporary ethnically based conceptions of 'national' identity, it must be understood in the historically specific context of earlier conceptions of Soviet nationhood. From the 1960 to the 1980s, Central Asian academics identified as part of a broader Soviet intelligentsia, members of a common social 'strata', albeit one internally stratified. They explained the success or failure of their work, particularly efforts to institutionalize sociology, within a broad, inter-republican framework (Alimova 1984; Leninizm i razvitie 1970; Skripkina 1983; Tabyshaliev 1984).

By late perestroika, however, this identity was challenged by exposure of discrepancies between the content and organization of work produced in the republics and that produced in the RSFSR (see, for example, Isaev 1991b). Emerging ethno-nationalisms and critiques of centre–periphery inequalities disarticulated 'republican' and 'national' identity in Soviet academe. Whereas Central Asian sociologists once criticized themselves for being too 'backward' to 'catch up' with their Russian colleagues, they now blamed disciplinary underdevelopment on their institutional and intellectual dependence on the Russian centre (Isaev and Bekturganov 1990; Isaev and Niyazov 1990; Vlasova 1989). They began to ask whether and how insights from a specifically 'Kirgiz sociology' could contribute to Marxist sociology on the whole, and questioned how self-knowledge had been proscribed by dependent development and 'democratic centralism' within the academy.

After independence, the ideal of an autonomous Kirgiz Soviet sociology was replaced with that of a new and fully independent 'Kyrgyz' form of knowledge (Baibosunov 1998; Bekturganov et al. 1994; Ryskulov 1998; Tishin 1998). 'Kyrgyzstan', it is argued, 'needs real, accurate and timely information' which sociologists argue only sociology can provide (Bakir Uluu 1994). The body of information published about sociology in Kyrgyzstan since independence in newspaper articles, research reports, theses and conference proceedings indeed gives off the impression that a new national sociology is emerging in the republic.

However, decades of initiatives to coordinate its institutionalization at the republican level have produced few tangible or sustainable results. The KSU laboratory, as an attempt to create an autonomous local base for social research in Soviet Kirgizia, is seen as an example of aborted development (Isaev and Bekturganov 1991; Nurova 2000). Plans to create an interdisciplinary Division of Social Sciences and Scientific Council on the Problems of International Development and National Relations in the Academy of Science during the 1980s did not, as proposed, facilitate networks between social science institutions or improve the quality of research on 'economic, sociological and legal problems that [had] practical national significance' (Koichuev 1988). And while *goszakazy* were hailed as the financial future of social research during perestroika, the withdrawal of state subsidies and revenue after independence and trends towards authoritarianism quickly drew a line under this alternative.

Just as Isaev and Bekturganov (1990) argued that poor coordination between academic social science and technological production demoralized researchers who saw few tangible results from their work during perestroika, those working in the post-Soviet period have made similar observations about the absence of 'mechanisms' for implementing scientifically informed national policy. Since 1990, Kyrgyzstani social scientists have continually repeated public appeals for improvements in education, research funding, legal support, journals and bulletins, an attestation commission for sociology and support for a professional association (Abdyrashev 1994; Baibosunov 1998; Bekturganov 1995a, 1997; BHU 1997b; Blum 1990; Isaev 1993a, 1998c; Isaev and Bekturganov 1991; Isaev and Niyazov 1990; Ryskulov 1998; Sydykova 1998). Ongoing attempts to found a national union of sociologists, which evolved from plans to create a Kirgiz branch of the SSA in 1990 (Sotsiologicheskoe obschestvennogo ob'edineniia 1999; Isaev 1991b, 2000), has been beset by financial difficulties and internal divisions amongst scholars themselves (Sagynbaeva 2003). The state, while still a source of ideological hope, has been a disappointment in reality. As one researcher put it, non-governmental organizations 'want to change things', but 'the [Kyrgyz] government has more power.... And they're not ready to get negative results'.

The rhetoric of national sociology is hence largely an intellectual, professional and political project rather than a descriptor. Sociologists have a sense that sociology is being institutionalized unevenly or even fragmenting; that, for example, 'sociology is developing in different directions' (Aldasheva 2003); that different institutions within the republic have incommensurable identities and functions (Omuraliev 2003), or that the 'separation' of professional sociologists 'contributes little to the creation and development of sociology' (Bekturganov *et al.* 1994; also Abdyrashev 1994). Even wide-reaching descriptions of trends in 'Kyrgyz sociology' obscure the fact each refers to trends in certain parts of the sociological community and excludes others.[7] In addition, there is considerable disagreement over the boundaries of the field in the post-Soviet context. Is sociology an indigenous form of knowledge or a 'Western' import? If the former, what are the distinguishing features of Kyrgyz sociology? If the latter, to what extent might it be generalizable to Central Asian society (Bekturganov *et al.* 1994; Isaev, Niyazov *et al.* 1994b)? Can sociologists integrate into the international community without sacrificing their particular national or regional identity (Isaev 2000)? Given the diversity within sociological theory, what schools of thought and methodological approaches may be considered legitimate contributions to the field in its new, post-Soviet form (Asanova 1995)? In the absence of authoritative decision-making bodies, who will be able to make these decisions? How will professional sociologists be trained and employed, and who has the power to certify the authority of their expertise? If sociology must renounce all affiliation 'illegitimate power', what are the proper boundaries between sociological work and power bases such as the state, media and international organizations? Finally, who, at any given moment, has the intellectual authority to define the legitimacy of any of these?

Rather than speaking of a 'national sociology', it is therefore more useful to propose that different groups of social scientists are engaged in producing different types of sociological knowledge in post-Soviet Kyrgyzstan. These sociologies share a number of similarities by virtue of their emergence in a common socio-political and economic context. However, one can speak of 'Central Asian sociology' only in the broadest of terms. After lecturing in both Kazakhstan and Uzbekistan, for example, Buckley (1999) noted the rising popularity of sociology and the general problems faced by all sociologists in the region. However, she also argued that 'the future for social science appears somewhat brighter' in Kazakhstan due to its less repressive government and a modicum of interest from the national Academy of Science. Furthermore, while sociologists from Uzbekistan and Kazakhstan consider themselves part of a common cultural space, many have also asserted the uniqueness of their own sociological traditions (Editor 1998; Luk'ianova 1990; Toschenko 1998). Within each of the republics, academics in different institutions practise different types of sociology, based on different sets of intellectual and professional traditions and relying on different institutional alliances within and beyond Central Asian society.

Reintroducing locality: the importance of institutional conditions

Despite the focus on the development of 'national' and 'international' sociology in Kyrgyzstan, local institutional contexts have therefore become increasingly important in the post-Soviet period. As academic institutions have become more atomized from one another and from the centre of the scientific world system, the definition and establishment of fields is particularly conditioned by institutional culture.

Comparative studies on national sociology provide insight into the political economy of social scientific knowledge in postcolonial societies, particularly how factors of colonial domination, submission and resistance impact upon knowledge production (Ake 1982; Akiwowo 1999; Alatas 2000a; Bujra 1994; Eisemon 1982; Filino 1990; Ganon 1965; Hiller 1979; Leoneri 1967; Rahman 1983). However, as this work focuses on explicating structures of academic dependency and the division of labour within the international scientific community, it tells little about how local and global social forces are experienced, interpreted and negotiated by social scientists in individual nation states.

Recent studies in the sociology of social science rather emphasize the importance of local institutional conditions in mediating the construction of scientific knowledge within national contexts (Camic 1995; Camic and Xie 1994; Small 1999). Small (1999), for example, argues that an emerging field of academic knowledge may be differently conceived and institutionalized under varying socio-institutional circumstances, even within a single national context. More comparatively, Bujra's (1994) historical comparison of government-led, inter-governmental, social-science-community-led and donor-community-led institutions in postcolonial Africa illustrates how differences in the ownership and function of social science institutions impacted the local development of the field. Similarly, Filino (1990) explored how different types of sociological institutions in Brazil and Argentina – Catholic universities, state universities, private teaching schools and independent research centres – were affected by the emergence of authoritarian regimes.

Decisions about the scope and content of sociology, its role in society, its relationship to other disciplines and practices and the relationship between teaching and research are made within departments that have very different philosophies, organizational cultures and access to material resources. In addition to examining the socio-political sources of such differentiation, we must also understand the effects of the narrower professional context in which a knowledge field emerges and explain how its conceptualizations are contingent upon departmental conditions. Gieryn (1983: 781) reminds us more generally that the demarcation of academic disciplines is 'routinely accomplished in practical, everyday settings: education administrators set up curricula that include chemistry but exclude alchemy; the National Science Foundation adopts standards to assure that some physicists but no psychics get funded; journal editors reject some manuscripts as unscientific'. He goes further to ask, 'how is the demarcation of science accomplished in these practical settings, far removed from the apparently futile attempt by scholars to decide what is essential and unique about science?' This question is particularly useful for analysing the development of sociology in the highly politicized atmosphere of Kyrgyzstan. Since independence, the local institutional context of sociological work has become a key factor in this process as scientific institutions in the republic diversify and stratify, and as new departments of sociology begin to emerge.

Despite the representation of Soviet sociology as a monolithic enterprise, institutions did matter under the Soviet regime.[8] However, until perestroika, the standardization of disciplinary knowledge was an integral part of Soviet science

(Bess 2000; Kodin 1996; Lisovskaia and Karpov 1999) and minimized the role of local institutions in the construction of social scientific knowledge. Sociologists working in various sectors of Soviet society shared a common intellectual and political culture, spoke one professional language and sought support and legitimacy from the same constituencies. While we have seen that certain individuals were able, within limits, to challenge intellectual orthodoxies and that they did not necessarily internalize official definitions of their work, the heteronomous position of the field within the Soviet power structure, the hegemonic politico-intellectual culture of the academy and the centralized organization of education and science prevented these variations from becoming sustainable alternatives.

Localized contexts of the production of academic knowledge became increasingly important during perestroika and after independence, as sociology became even more radically decentralized (Dunston 1992; Kerr 1992; Sutherland 1992). Despite the sense of professional anomie that this created, it was also viewed as an opportunity for sociology to recover from the 'deviation' of the Soviet experience and resume its 'natural' course of development as an autonomous scientific discipline. Many Kyrgyzstani sociologists believed that the emergence of greater intellectual freedom would automatically stimulate the institutional development of social science in one direction: towards non-Marxist, 'world sociology' (Isaev 1993; Isaev, Niyazov et al. 1994b). Many are sceptical of arguments that this process can or should be diverse. Approaches that do not fit into the consensus about what constitutes the 'correct' path of development are perceived as new forms of politicized deviation. This has made it difficult for many sociologists to come to terms with how and why multiple conceptions of sociology have emerged in Kyrgyzstan, and hence to overlook the significance of local institutional conditions embedded in more national and global relations.

The experiences of Central Asian academics working to reconceptualize and re-institutionalize social science after the collapse of Soviet communism contribute to a more nuanced, slightly more tragic narrative of social 'independence' – one in which movements towards freedom are also movements towards new types of dependence, where the dismantling of one set of relationships of domination facilitates the construction of diverse others, and where processes of postcolonial identification and autonomization are embedded within broader relations of neo-colonial dependency. What has remained constant in the face of all this disruption, however, is the desire to develop a body of knowledge that will be scientifically sound, politically useful and socially authoritative, and the development of strategies for navigating competing epistemological logics in the attempt. From the perspective of the sociology of knowledge, this is to be expected. A nation's political independence – whether ascribed as consequence of imperial collapse of empire or established as a result of anti-colonial struggle – constitutes a new context for a new politics of knowledge within a society, rather than the end of such contexts altogether; the site of new articulations between organized knowledge and socio-political power.

6

RE-DISCIPLINING KNOWLEDGE IN KYRGYZSTAN

Alternative visions of sociology between Marx and the market

Political independence from the Soviet Union failed to herald in Central Asia the dawning of a new era of knowledge production in which both objective truth and its social authority advance progressively, cumulatively and in accordance with the logic of science. This came as some surprise to those who theorized the politics of social scientific knowledge as an artefact of socialized totalitarianism; others had predicted that the intellectual and institutional legacies of this system would create path dependencies that hinder development in the post-Soviet period. However, while the phenomena concerning contemporary Central Asian social scientists (struggles to demarcate the boundaries of legitimate knowledge where orthodoxies of truth have been dismantled and to realign their own relationships with forces of social power, contests for intellectual and professional authority within fields that conflate 'scientific' and 'political' rhetoric, and competition to garner financial and social support for institutional survival) have taken on new forms, the phenomena themselves are not particular to post-Soviet nationalization or even decolonization on the whole. They are rather indications of 'science as usual', of the 'normal' social processes through which fields of knowledge are constructed and deconstructed or legitimated and debunked, particularly in the form of academic disciplines (Fuller 1993; Gieryn 1983; Gumport and Snydman 2002; Latour 1987; Shumway and Messer-Davidow 1991).

Central Asian academics are therefore creating the meaning of organized knowledge the post-Soviet 'transition', not by fulfilling a monolithic and teleological process, but by negotiating epistemological histories with existential conditions and value orientations, all within the context of changes in the global relations of science. This process is particularly visible in the everyday practices of sociologists working to create, develop and reform university departments in the region. The following sections explore the post-Soviet history of two sociology departments in Kyrgyzstan, one at the Bishkek Humanitarian University (BHU) and the other at the American University–Central Asia (AUCA). They are similar in so far as both consider themselves to be leading departments in the field; both also emerged as part of the post-independence growth in the social sciences and share a common project of institutionalization. Both are expected to design and stabilize programmes for awarding academic degrees with extremely limited time and material and cultural resources (Ablezova 2003; Sagynbaeva 2003). The demand for 'instant institutions' stems in part from developmentalist discourses of institution building (Sakwa 1999), but also from sociologists' own theories about the conditions for institutionalizing a 'mature' version of social science. At these levels, the departments share a common socio-historical experience.

However, they differ dramatically in the way that academics conceptualize and practice sociology. Since its establishment in 1993, the BHU department has built a reputation as the country's leading state sociology institution, cultivating an identity as an applied national profession, which contributes to state-sanctioned social reform through empirical research. The AUCA department, established later in 1998, is considered the principal private and 'international' sociology institution, and encourages the development of sociology as a liberal art, form of social criticism and semi-commercial enterprise. These differences are often interpreted as manifestations of a political struggle between 'Soviet' and 'western' philosophies of knowledge, between academic conservatives protecting the past and an intellectual vanguard of the future. Such a discourse certainly frames competitive sentiments between the two universities, which seldom interact even though individual academics sometimes work in both. Those working exclusively in the state system, for example, tend not to recognize AUCA as a legitimate university and therefore know little about its sociology. Similarly, the 'Soviet' heritage of BHU keeps those privately employed at AUCA, whose professional identity rests in part on their own novelty, from affording it serious consideration. The founder of its sociology department, in fact, says that she chose to pursue her vision in AUCA because it was less 'Soviet' than either BHU or KNU.

The terms 'Soviet' and 'independent' in this case are reductivist referents for complex sets of conditions within which knowledge is currently being produced and reproduced. It is important to understand how the 're-disciplining' of sociology in the post-Soviet period is related to structural matters such as each department's institutional history, sources of funding and social capital, and relationship to state and society. Other factors, such as culturally specific ideologies of science and its role in society, the emergence and affiliation with or distance from new scientific 'centres' in the West, and the diverse intellectual orientations of individual academics also influence how the field is constructed. Ultimately, however, each of these factors is mediated through academics' decisions about how to negotiate the competing logics and demands of Marxism, capitalism and institutional professionalism in social scientific practice. The experiences of academics working in AUCA and BHU suggest that, instead of being engaged in a 'transition' from one to the other, each is strategically reconceptualized for particular purposes. The following sections examine this process and illustrate the contingent nature of disciplinary knowledge and practice in Kyrgyzstani sociology.

94

An applied science for social administration: sociology at the Bishkek Humanitarian University

Founding, funding and faculty at BHU

The establishment of sociology at the Bishkek Humanitarian University (BHU) was no isolated event. In 1988, the Communist Party had decreed that universities should begin to systematically create such departments (Aldasheva 2003), and from 1989 to 1993, Isaev and his associates had been engaged in constructing a new 'sociological' identity for the department of Sociology and Engineering Psychology, located in the Frunze Polytechnic Institute (FPI) (Blum 1990; Ismailova 1995; Osmonalieva 1995; Sydykova 1998).[1] In 1993, Isaev was invited to chair a newly created Faculty of Socio-Political Sciences at BHU and established a new laboratory, transferring records, projects and staff from the FPI (Baibosunov 1998; Isaev 1999a; Osmonalieva 1995).

Despite successive changes in leadership since that time, the department has protected its reputation as a national institution created by and for the Kyrgyz state and operating as a service to the nation by training new scientific elites and making contributions to governmental administration in the form of 'scientifically grounded recommendations' for social policy (BHU 1996) – even as it has become increasingly evident that the state itself has little regard for social research (Sydykova 1998). While academics' main aim is to train elites to fulfil administrative functions for the Kyrgyz state, they aim to be allied equally with state and society and therefore assumes a unified, pro-government front in its official activities while permitting more subtle types of criticism within (BHU 2002b, 2003a).

Sociology at BHU is both integrated into the state system of higher education and embedded in a new and competitive educational 'market', and thus exemplifies broader tensions within higher education in the republic. On the one hand, the university is funded by the Kyrgyz state, politically supervised by the Ministry of Education and institutionally subordinate to the ruling government (BHU 1995). On the other hand, the state investment in education has steadily declined since independence. Sociologists feel this acutely as salaries decline, Russian translations of American and European work remain unattainable and academics cannot afford to publish in Russian journals (Aldasheva 2003; Asanbekov 2003). One admitted feeling that while the department was 'not a priority for anyone. The money goes to wealthy universities. In general, this department gets nothing'. As a result, many academics have grown increasingly reliant on foreign sources of funding such as the Soros and MacArthur foundations. By 1998, in fact, the department encouraged its instructors to seek external funding for both teaching and research. Thus, although dependent on the state, they became increasingly 'prepared for the necessity of doing fundraising for the introduction of new courses and improving the material base of the faculty' (BHU 1998).

This, however, creates other tensions within the department. Such grants, when they are received, offer sociologists little autonomy and are often conditional, many being commercial contracts commissioning sociologists to gather data for foreign clients. One researcher remarked that 'when a foreign firm invites us [to do research], we do not know the results.... We have the data, we interview everyone, but...the firm does the analysis itself. And we don't even have a publication of this here'. There are sobering stories of sociologists from other institutions undertaking legal battles to defend their rights and reputations against more powerful foreign grant-giving organizations (Sotsiologicheskoe obschestvo Kyrgyzstana [n.d.] and Skorodumova 1998). Despite his criticism of the colonial nature of such relationships, Isaev (1993) has even suggested that under such stringent circumstances sociological laboratories should be used for 'fulfilling *zakazy* from the state, private or other types of organizations and enterprises on a khozgovorniy level [financial contracts], which is an important source of additional salary for teachers, co-workers and students'. Such research generally addresses topics of interest to these agencies, such as migration, business, unemployment and reproductive health; however, because it frequently takes the form of empirical data gathering for use in specific policy reports it has little impact on the department's research or teaching programme on the whole.[2]

The conceptualization of sociology within the BHU department has additional cultural foundations. Faculty responsibilities include not only student mentoring and heavy amounts of bureaucratic recordkeeping, but also conforming to professional norms within the department. Generation matters; as do ethnicity and gender. Traditions of academic hierarchy are pronounced within the department, which employs almost exclusively 'local' instructors, a number which belong to what Ibraeva (2003) calls the 'older' and 'intermediate' generations of Soviet-trained sociologists.[3] She bases these categories not only on chronological age, but on 'spirit of thinking' as well. In her view, members of the 'older' generation continue to value the traditional professional hierarchies which are being delegitimized; they have 'experienced the shock of being unneeded' and, as they are 'not particularly familiar with the sociological method of research and are not in a position to answer the demands of the time in the face of growing competition', are currently 'living through a dramatic situation'. Some older faculty members do strive to adapt to the 'new conditions' by incorporating non-Marxist theories into their research and teaching and expanding their repertoires of research methods and pedagogical techniques. However, within the department, they still value and impose the traditional professional hierarchies in which they continue to hold superior positions.

Members of the 'intermediate' generation of sociologists simultaneously reinforce and challenge these divided relations. By Ibraeva's definition, these are middle-aged academics who retrained as sociologists after independence and for whom it is less daunting to 'meet contemporary demands' (i.e. conform to hegemonic norms of knowledge production), in particular by conducting 'Western-style' research and adhering to international norms of academic etiquette in public forums. This group, however, is also facing a 'dramatic period, in so far as for many people it is sustained by traditions (reverence for elders, hierarchy in the professional sphere, the aim of exploiting young specialists, and etc.)' (Ibraeva 2003). Many of this generation are intellectually frustrated, for while they may be inspired by new schools of thought and possibilities for further education, they generally lack the linguistic skills needed to take advantage of foreign-sponsored opportunities and are sometimes excluded and discriminated against by foreign organizations.[4] Thus, while they may be intellectually open and eager to participate in educational initiatives, they are constrained by what they see as legitimate or at least necessary norms and challenged by younger academics who have been able to acquire more 'marketable' professional skills.

The youngest generation of sociologists, predominantly female, is the greatest beneficiary of new, often foreign-led initiatives to retrain social scientists in the post-Soviet period. However, they are also most subjected to exploitation and are often targets of professional envy (Omurkulova 2003). Foreign organizations often prioritize younger instructors as 'mediators' that can disseminate and popularize new ideas within their home institutions; however, they often receive little encouragement or opportunity to do so. Deference to older, sometimes even less qualified colleagues severely constrains younger instructors' academic potential and ability to initiate intellectual exchange within their own departments. Intellectual and organizational change in the definition of sociology, teaching and research, and professional norms continue to originate mainly from above, and those working at lower rank tend to perpetuate this relationship out of fear, apathy or professional ambition.

These social hierarchies reinforce a tendency toward authoritative knowledge reproduction rather than creative knowledge production in both teaching and research. However, while this largely eliminates ambiguities about authority within the department, it does not prevent power struggles among faculty members. Beyond the day-to-day departmental relations, all are equalized on the wider and increasingly significant 'market' for contracts and grants; in fact, younger faculty are often advantaged by their higher levels of social capital in the post- and non-Soviet science system, as well as by ageism in grant-giving agencies. This engenders competitive rather than cooperative relations among the faculty, with academics – not unlike scientific researchers elsewhere (Rothman 1972) – hoarding ideas and information in order to protect their jobs, niches of expertise and edge in extra-departmental opportunities.

This also affects activities such as curriculum development, which is ostensibly a collective responsibility. Instructors debate the merit and appropriateness of new textbooks before introducing them into the curriculum (BHU 2001, 2002a) and all must present their lectures and lesson plans to the department for approval. In reality, however, curriculum development a power-laden and bureaucratic process, dominated by senior members of the department who have considerably more decision-making power than their younger colleagues (although, as will be discussed later, even this influence is constrained by external political and economic forces). Because new courses (*avtorskie kursy* or authors' courses) are considered scholarly work (even if they sometimes are not original), academics regard teaching materials as private intellectual property and are reluctant to share them with others. As a senior sociologist explained, contrary to the Soviet period in which she claims they 'all helped each other', she feels that academic life has become highly competitive. This atmosphere is fostered partly by the current individualized strategies of grant-giving, partly by the decline of state funding, and partly because while sociologists now have relatively more opportunity to be intellectually creative, they have less time and fewer resources with which to do so, and decreasing confidence in the moral standards of academic integrity.

In other words, while the department outwardly advocates its commitment to the advancement of social scientific knowledge and traditional ethos of science, the internal organization and culture instead create conditions for its reproduction and the creation of counter-norms. However, this also dovetails with the department's objective to establish authority as producer of the nation's new scientific elite. Sociology is represented as a standardized body of knowledge and skills that can be transmitted with precision from one generation to the next and applied scientifically to predict, prevent and alleviate a range of social problems. While portrayed as a new model of post-Soviet social science, it retains elements of the state-oriented, applied-professional model of academic social science that was ascendant during the late Soviet period. The concepts and epistemological foundations of Soviet sociology, rather than disappearing, have been modified to align with new positivist and non-Marxist discourses about the nature, purpose and politics of knowledge.

Constructing sociology: teaching at BHU

The mission of the BHU Sociology Department was unambiguous from its inception: it was created to 'prepare a new generation of cadres to administer collectives, regions and states' (BHU 1994); to train cohorts of specialists to contribute to efficient governance. As during perestroika, this implied not only responding to specific questions posed by members of the political establishment, but also identifying social problems, thereby ostensibly hastening social reform. Supporting this is a philosophy of knowledge that separates the field of sociology into three distinct components: 'general theory' (historical materialism), 'particular theory' and 'applied sociology' (Nurova and Shaimergenova 2000: 4; Vucinich 1974). This tripartite model of the discipline, which had emerged during the 1980s, is viewed as a linear progression with general theory being the epistemological foundation and applied research the ultimate responsibility of a professional sociologist (see Kabyscha 1990). A recent departmental brochure makes the application of this philosophy explicit: 'having *received the profession* of sociology, our graduates will have the opportunity to, at a professional level, conduct sociological research and do scientific analyses of social phenomena and processes and, on the basis of these, make prognoses and recommendations' (BHU 2003, italics mine).

The production of 'trained cadres' is therefore seen as the central aim of sociology's institutionalization in both education and research (Isaev 1993, 2000). This is evidenced by the way in which the department sets priorities for its learning outcomes. By 2003, students choose to study in one of five professional specializations: 'sociologist–economist', 'sociologist–marketing specialist', 'sociologist specializing in computer technology', 'sociologist–legal specialist' or 'sociologist–instructor of social sciences' (BHU 2003). In addition to structuring the curriculum, these categories also buoy the status of sociology by extending its authority into 'prestigious' fields such as economics, marketing, computer technology, law and education. By affiliating it with these fields and creating 'sociological' specializations in each, the department can offer students a unique 'product' that is also predictably marketable through its affiliation with more 'stable' degree programmes.

This applied-professional orientation is reinforced by faculty members' beliefs about the role of sociology. Isaev, for example, had advocated professional training for sociologists long before he assumed the first chair of the faculty (Sydykova 1998). As early as 1990, he criticized the Soviet state for failing to take this seriously and recommended introducing courses in the sociology of industry, work and administration into the new curriculum for perestroika (Isaev 1991b, 1993a; Isaev and Bekturganov 1990). The agenda also resonates with the intellectual convictions of other faculty members in the department. One, for example, asserts that

> [t]he role of sociology is to study social reality. A sociologist must know this reality... must study what is happening in society, analyze it, and say what's wrong with the social mechanisms and what can be done to alleviate the problems. They must also give advice about what needs to be done to cure social illnesses or make it so that they do not emerge.
>
> (Anonymous)

Another also prioritizes the 'study of reality', arguing that sociology has a 'great role' to play in helping Kyrgyzstani society to recover from its 'totalitarian past'. Citing Tishin's (1998) 'twelve functions of sociology', she notes that the study of social reality in all its variety can play an ideological as well as a technical role by forcing students to remove the 'rose-coloured glasses' through which they often understand society.[5]

Towards this end, the department has developed a core of courses or 'disciplines' deemed necessary for professional expertise in sociology.[6] Its authors, as in other Central Asian republics (De Young and Valyayeva 1997), draw inspiration from experience in Soviet universities, contemporary curricula from Moscow and St Petersburg, and the national standards for sociology, themselves based on a Russian model (Aldasheva 2003; Ministry of Education 1994;

Ryskulueva 2003). The courses reflect both the influence of state standards and the existing knowledge base of the faculty. The broad core of survey courses is supplemented by 'general professional disciplines in the subject' geared towards professional (Ministry of Education 1994: 13). Given the dearth of qualified sociology instructors in the region, the department has capitalized on the experience of its existing faculty. Their introduction is thus less systematic than that of core fields and tends to depend on overall student demand (Aldasheva 2003).[7]

This combination of top-down and ad-hoc growth has created certain challenges in curriculum development, including low levels of motivation and professional commitment. Under the conditions, neither junior nor senior faculty have much say in determining the range of courses they teach. This lack of autonomy has bred a lack of imagination; innovation is theoretically encouraged, but even ideas that are marginally unfamiliar are often rejected. Furthermore, the breadth of the curriculum requires each instructor to teach between two and six courses per year, with less experienced lecturers often teaching heavier loads.[8] One senior lecturer reflects on how this affected her teaching:

> Unfortunately, until recently, teaching...was linked with an incredible teaching load and the need to develop the most various courses. Thus, the courses in my pedagogical toolbox were quite diverse: gender sociology, sociology of mass media and mass communication, urban sociology, political sociology, history of sociology, introduction to sociology, sociology of management and sociology of conflict. The practice of 'plugging up' gaps in the instructional programmes at higher education institutions has terrible consequences, [such as] poor quality of courses, particularly in connection with the lack of literature and other instructional materials.
>
> (Anonymous)

The 'plugging up' metaphor encapsulates the current philosophy and practice of curriculum development in sociology in Kyrgyzstan – not only in BHU, but in many state and private institutions throughout the republic. There are empirical discrepancies between intellectual expectations, educational and political requirements and the human and material resources available to fulfil both. Sociologists working at private institutions with curricula that diverge from both the national standards and traditional models of Soviet education also struggle to balance competing demands for programmes that satisfy both local and international conditions with the need to promote quality teaching and research.

Curricula for undergraduate sociology are also shaped by epistemological assumptions about what constitutes legitimate sociological knowledge and how and why people acquire, produce and reproduce it (Gumport and Snydman 2002; Popkewitz 1991). The inclusion or exclusion of theoretical knowledge, methods and foreign or indigenous work thus reflects tacit beliefs about the nature and role of knowledge and the learning process itself (Lisovskaia and Karpov 1999). In developing courses, instructors at BHU rely heavily on national standards elaborated by a special disciplinary committee under the Ministry of Education, Science and Culture in 1994 and revised ten years later (Ministry of Education 1994, 2004). The standards outline in considerable detail an applied-professional model of the field. Sociology education must be organized to

> help industries, institutions, organizations, commercial structures and legal and physical individuals to expose and resolve social problems . . . Successful graduates from state institutions must be able to inform reform, assist in administration and predict future social trends in areas such as health and welfare. They may also teach in scientific and educational institutions, commercial enterprises, and research or public opinion centers.
>
> (Ministry of Education 1994; see also BHU 1997)

Whereas prior to independence intellectual standardization was driven by the party's need to maintain political control through intellectual hegemony and to rationalize a system of mass higher education, it became increasingly integrated into capitalist labour economics and the state's desire to retain control over the content and method of instructional process in a deregulating educational arena. According to Ryskulueva, a specialist at the Ministry of Education, there are two possible explanations for why the ministry introduced disciplinary standards instead to replace the older practice of issuing formal *uchebnye plany* [instructional plans].[9] First, she argues that national standards give students 'flexibility' and the 'right of mobility' in an educational system that has become governed by student demand as must as state planning.

> If we don't have any standards then [a student] will have trouble transferring [between universities]. In one [institution] they will teach him according to their own programme, and he will have problems in another. . . . If you have this kind of difference, you can't continue to educate students because the programmes are entirely different. Therefore, we introduced the standards. Standards give students the possibility to realize their right of mobility. We work out standards that everyone must follow in the regions and in Bishkek. If everyone has these requirements, there won't be any problems.
>
> (Ryskulueva 2003)

Because the state no longer has the financial means or authority to oversee all activities within a proliferating educational system, the standards also have more overtly disciplinary functions. These include the regulation and containment of academic freedom, minimization of diversity in new educational philosophies, and prevention of the disintegration of historical relationships between education and labour. To quote Ryskulueva again,

> [a]cademic freedom is increasing gradually. We cannot just give freedom straight away because [instructors] are not yet used to it. We defined the correct contents of the curriculum for how many years, one hundred years now? People studying in every institution and university knew what, how many and which disciplines they needed to study in order to meet the requirements. But... we don't have this yet. Our higher education institutions have absolutely no idea. First of all, they accept all students who come to them but don't consider whether their graduates will be able to get work. It is important that students come and pay money, and that's it. They give it and then everything falls apart.
>
> (Ryskulueva 2003)

From the ministry's perspective, the decentralization of education and university-based job training programmes threatens social and economic stability. Therefore, the main task set before sociologists working in state institutions after independence was not to innovate, but to recreate institutional frameworks for standardization to replace those which had either collapsed or been abandoned as undesirable.

As employees of a state institution who are dependent upon the goodwill of the government for professional security, the department elevates compliance with the national standards to a matter of disciplinary identity. University degrees are still conferred by the Ministry of Education and not individual institutions; students not holding a Kyrgyz diploma are often seen to have graduated from 'inferior' schools and thus are at a disadvantage in the national labour market.[10] While the national standards are technically 'suggestions' rather than require-ments as were Soviet instructional plans, the extent to which a department's curriculum conforms to them has a direct influence on whether it is granted attes-tation from the Ministry of Education. Offering a state-sanctioned curriculum not only enables the department to produce 'employable' students, but also bestows upon its degree a type of contingent political legitimacy not afforded to departments in private institutions.[11]

Disciplinarity

This also influences the boundaries of the field itself, including its relationship to other social sciences and to the society with which it seeks to be engaged. 'Disciplinarity', or the process by which knowledge units are constructed, altered and deconstructed, has been an emerging topic of interest since the 1980s (see Messer-Davidow et al. 1996; Good 2000; Lemaine et al. 1976; Shumway and Messer-Davidow 1991). The term 'discipline' has a dual meaning, referring to both the intellectual boundaries of a knowledge unit and the practices through which these boundaries order or 'discipline' thinking and action in that sphere.

Sociology departments in Kyrgyzstan have different notions of their own disciplinarity. In some cases, these boundaries are fluid and contested. At BHU, disciplinary boundaries are fixed and theorized as appropriate and desirable. Sociology is defined as an empirical object, a naturally occurring body of knowledge possessing a coherent history and stable set of characteristics that transcend time and space, and a universal standard against which inferior classes of social scientific and lay knowledge can be measured. Reflecting the common *doxa*, for example, Isaev (2003) argues that 'as a science, profession and subject, sociology has no less than a two-hundred-year-old tradition of development'; that it is 'studied in nearly all higher education institutions in the civilized countries of the world'; and that 'on the eve of the twenty-first century, a single world sociological science has been formed and objectively exists'.

The department adopted the Kyrgyz Ministry of Education's classification, G.12(521200) 'Sociology', to distinguish sets of knowledge and skills from those allocated to other professional specializations such as politology, social work, psychology and pedagogy. In this framework, sociology addresses only those matters that are deemed to belong to the 'social': migration, ethnicity, gender relations, and social change and stability, as opposed to more 'political' matters, such as government, political parties and elections, which are the purview of politology. For the ministry, these distinctions not only establish neat intellectual categories but also enable potential employers to hire specialized graduates for positions requiring particular skills. This is perceived as a critical 'selling point' of sociology in the new educational market. Although the department offers introductory sociology courses for students of other departments, its primary agenda is to distinguish it as a discrete discipline. The effort to erect unambiguous rhetorical and institutional boundaries between sociology and other fields is also motivated by a desire to promote sociology as a unique way of knowing about society in order to expand sociologists' authority and right to resources.

Therefore, in addition to demarcating the knowledge and skills that differentiate fields, sociologists are also particularly sensitive about distinctions between social science, politics and lay knowledge. The well-educated professional sociologist must be qualitatively different both from politicians, who it is argued are prone to intentionally distort social reality, and from members of the general public, who are seen as lacking the necessary information and skills to apprehend social reality accurately. Sociologists are portrayed as scientific guardians of social consciousness, whose authority stems from possession of specialized disciplinary knowledge and the ability to conduct scientific studies that produce objective representations of social reality.

However, there are tensions between the quest for scientific legitimacy and that for social relevance in sociology. The department has not resolved the discrepancy between its overtly political relationship to the state and its insistence that the knowledge and practice that it purveys are essentially apolitical. Boundary-work is thus also used to distinguish the epistemological nature of sociological knowl-edge (i.e. 'scientific') from its social and political application. On the one hand, faculty members are politically responsible for producing knowledge contributing to the creation, maintenance or restoration of social order; overt challenges to the ruling regime made at the department level, for example, could result in reprisals. Formal teaching and research therefore reinforce the notion that social change is effected through advising political and managerial elites who can then more 'scientifically' design and implement social reforms from the top-down. On the other hand, however, sociologists maintain that one of their main roles is to 'expose' social reality and unpleasant truths about social life in Kyrgyzstan. As such, informal classroom teaching sometimes has a more critical edge. Students may be encouraged to think critically about issues such as the legality of elections, political participation or the meaning of national culture; however, the work they produce for examinations still fits comfortably within the bounds of the standards for professional knowledge outlined by the Ministry of Education. In order to balance these competing roles, the department emphasizes an ideal of scientific politics and asserts that a truly 'scientific sociology' can contribute to state-sanctioned social reform through both scientific skills and social criticism. By claiming to offer a window onto objective reality through providing students and the public with objective knowledge, sociologists can also promote the restoration of social stability in an anomic society. The production of objective truth about society is therefore adopted as the primary discourse surrounding sociological research at BHU.

Research

The relationship between teaching and research in the department is based on a model previously advocated during perestroika, which asserts that all sociology departments should be affiliated with a research laboratory. During the late 1980s, however, combining teaching and research in Soviet universities was a radical rethinking of the relationship between education and science, which hitherto had been regarded as fundamentally different social institutions. Social research was conducted primarily for the Soviet politico-industrial-military complex and produced in 'scientific' institutions such as the Academy of Science and on-site research centres. Social science education, on the other hand, was carried out to provide scientific institutions with a skilled labour force, and was situated in specialized secondary and higher education institutions such as universities and technical institutes.

The theoretical basis for this model was that there should be a 'dialectical relationship' between 'two types of scientific knowledge' – material knowledge on the one hand and 'spiritual' knowledge, or thought, values and imagination on the other (Isaev 1993). This required that 'a sociologist–instructor must be above all a highly qualified specialist and [able to] combine a high theoretical level of sociological knowledge with the talents and skills of conducting concrete empirical research' (Isaev 1993). A recently published sociology textbook criticized pedagogical materials because their authors 'do not do concrete sociological research' and therefore produce books which are, on the whole, merely traditional or classical 'compilations of sociological views' (Tishin 1998: 3).

Within the BHU department, there have been a number of initiatives to include students in research work.[12] Nevertheless, the dialectics of teaching and research remain part of a formal rather than substantive project. While students are required to assist with projects carried out by academics and conduct individual research projects, apprenticeships are often formulaic and not systematically integrated into students' learning experiences. Instead of being dialectically related, it is perhaps more accurate to say that sociological teaching and research co-exist, but as discrete activities. As with the curriculum, research at BHU is highly structured, centrally organized and dominated by a small number of senior academics and defined as a departmental rather than individual activity. Although some pursue independent work outside the confines of the working day, one noted that 'very little time, practically none, is left for working on any sort of scientific problem'. All research must conform and contribute to the department's 'general scientific theme' (BHU 2000).[13] Each year, this broad programme is divided into a number of narrower themes, each one investigated by a 'team' or designated group of instructors supervised by a senior faculty member (Nurova 2003). Early themes included the development of new social groups in the process of transition to market relations,[14] the particularities of the creation of new political, military and economic elites in the conditions of democracy,[15] and changes in the process of transformation (BHU 1994, 1995). From 1996 to 2000, this was expanded to include other interests: socio-cultural processes (e.g. globalization, migration, mass media and ideology), labour and distributive relations (e.g. unemployment, internal migration and poverty), changes in social groups (including migrants and women), deviant behaviour and interethnic relations (BHU 1997).

The research conducted by the programme's postgraduate students (*aspirants* pursuing candidate degrees) reflects the department's insistence on both professional hierarchy and intellectual homogeneity. In 1994–95, for example, seven were writing dissertations on the 'development of social groups', two on 'problems of the political elite' and two more on general topics of 'social development in Kyrgyzstan'. Others later focused on democratization, stratification, values, religion and deviant behaviour; in 1998 some began to study 'civil society', and by 2001 the department established a new thematic component on gender (BHU 2002).

Postgraduates have limited opportunity to define their own topics of research, and many are assigned studies that correspond to the work of a senior academic and the department's 'general scientific theme'. One, for example, met with considerable resistance when she proposed to write a masters thesis on narcotics, a subject which she became interested in while studying abroad:

> I wanted to write my dissertation on narcotics, but they didn't let me. They said, 'oh you're such a girl, you still don't have enough information, it's an overly dangerous theme'. Hence, though I wanted to do research about drugs, I don't know why, I couldn't.... I wanted to work with

[another supervisor], but he didn't want to give me to anyone because he knew I would do all right. I went to him seven times and told him I didn't want to do this theme... I wanted to write about narcotics.

(Anonymous)

The homogenization of research interests, however, reflects more than an affinity for intellectual domination. It is exacerbated and at times seemingly necessitated by the shortage of scholarly competencies and materials in other areas (BHU 1995). The national library has few academic resources on contemporary sociology, the university library stocks only a few books on sociology and selected volumes of sociological journals, and the smaller library maintained by the faculty houses mainly theoretical and introductory textbooks (Isaev 2003a; Osmonov 2001).

Boundary-work and contingency in sociology at BHU

Decisions about curriculum, research, disciplinarity and professional identity made within the BHU Sociology Department are excellent examples of a political economy of knowledge 'in transition'. Everyday academic practices legitimize and reproduce a particular discourse of post-Soviet social change: that of a class of professional social scientists sovereign and democratizing nation breaking fetters of totalitarian communism and building cultural institutions of capitalist democracy. Sociologists, defined as specialists in a new science for the new society, are characterized as midwives to this process – their extra-social, apolitical and 'scientific' status distinguishes them from their Soviet predecessors (and in many cases, from their former selves) who have been reclassified as 'pseudoscientists' or intellectual *apparatchiks*. While they remain subordinate to the state, it is represented as reformed – power that seeks to base decisions about human welfare on scientific research; one that values truth over power; a state in which sociologists can finally realize the professional influence and identity they sought to establish under Soviet communism.

In practice, however, professional practices embody the more complex struggles and tensions of a non-democratic, postcolonial state. While academics continue to legitimize state-centric theories of the relationship between science and society, economic crisis, decentralization and marketization of higher education, and authoritarian national politics compel them to establish alternative alliances with non-state constituencies such as international organizations and create spaces of relative, often individualized, criticism within the institution. Beholden to the government for political legitimacy and basic subsistence, drawing primarily on experience accumulated in the Soviet academy, they are also expected to achieve fluency in new theoretical and methodological principles and to embrace new types of professional ethics. Traditional academic hierarchies are simultaneously maintained and dismantled as 'older generation' sociologists reproduce their power within the university and those in the 'younger' generations contest it without. Here, the 'transition' of sociology is fraught with uncertainty. Sociologists, engaged in reconstructing the intellectual and political identity of the field, do so while strategically negotiating alliances with a bankrupt and authoritarian state, the unstable and uneven relationships ensuing from dependence on development organizations, or reliance on commercial funding.

To explore how the same structural contingencies took different form and coalesced with different subjectivities to cultivate the emergence of an alternative conceptualization of sociology, we now turn to the development of the field at the American University–Central Asia.

Between scholarship and service: sociology at the American University–Central Asia

Founding, funding and faculty at AUCA

The Sociology Department at the American University–Central Asia (AUCA) was founded in 1998 when the university was five years old. It was established in 1993 as the Kyrgyz–American Faculty, a small department housed in the English-language faculty of KNU. Its founder, a charismatic and politically influential language teacher named Kamila Sharshekeeva, aimed to train students in English and introduce them to 'market-oriented' fields such as business administration, law and economics. In 1997, the school separated from the National University, strengthened its formal ties with the US government, changed its name to the American University in Kyrgyzstan (AUK), moved into the former headquarters of the Kirgiz Republic's Communist Party Supreme Soviet and was conferred independent status by presidential decree (AUK 2002; Ministry of Education 2000).[16]

At this point, the institution began a rapid transition from a small, professionally oriented Soviet faculty to an American liberal-arts-style private college. The shift included a reorganization of the disciplines, in particular a new focus on building departments of social science (Reeves 2003). This university-wide agenda coincided with an initiative to establish a 'new kind' of sociology department in Kyrgyzstan, led by Ainoura Sagynbaeva, a Moscow-educated sociologist who at the time taught short courses in sociology. In 1997, she was inspired to expand the university's offerings in sociology while on an academic exchange programme to the US. There, she encountered a range of sociological perspectives that were unknown in Kyrgyzstan, a strong scholarly community and specialized degrees in which students made choices about how to design their own educational programmes. 'To be honest', says Sagynbaeva, 'when I told the students that I dreamt of a school, I didn't mean to have a department. I simply meant colleagues who would understand me, who would love sociology' (Sagynbaeva 2003). Sagynbaeva believed that AUK was the most suitable site for her project because it was 'experimental' and less 'Soviet' than either BHU or KNU.

The university enjoys both symbolic prestige and material privilege for a variety of reasons. Owing to its relative autonomy from the state and its affiliation with the US government, it became endowed with almost millennial status as the regional standard-bearer of 'modern' educational reform and 'internationalization'. It also attracts large numbers of wealthy, high-performing students who were also often recipients of educational grants sponsored by US organizations. Hence, both the administration and student body have had interests in the Westernization, specifically the Americanization, of the university. Finally, as Reeves (2003: 28) points out, the school had no Soviet identity to 'shake off'. This novelty factor means that institutions such as AUK, 'whilst facing considerable difficulties of their own (notably, establishing themselves as "reputable" in the eyes of longer-established institutions)... have far greater leeway to introduce reforms without this being seen as revolutionary'.

However, in the context of post-Soviet nationalization, political autonomy *vis-à-vis* the national state and close cultural and economic connections with 'the West' have also made AUCA an open site of political and ideological struggle. In fact, there have been a number of attempts to close the university, curtail its experimental activities, and force the administration to conform to more traditional types of educational management. Resistance to its organizational mission has emanated from within the university as well as externally, particularly from instructors who wanted to maintain a more specialized and traditional pedagogical philosophy. There has also been some resistance to 'internationalizing' education in a country that is self-consciously constructing a 'national' identity (Reeves 2002). Finally, many of the university's large-scale administrative initiatives, such as awarding merit-based admission and scholarships, charging high tuition fees, penalizing bribery and corruption, encouraging cross-curricular thinking and adopting an American-style 'credit-hour' system of registration, have been criticized by other members of the educational community, many of whom feel threatened both personally and professionally by these changes. Among other things, AUCA is a

> reminder of the glaring inequalities that have polarized Kyrgyzstani society for the last ten years and the unfettered penetration of the market into areas of social life, education among them, that were previously free of such logics. As such, it is often seen as representing a set of values and an educational philosophy rooted in liberal individualism that is alien to, and inappropriate to meet the needs of, contemporary Kyrgyzstan.
>
> (Reeves 2002: 22)

Emerging within this political and cultural milieu, the Sociology Department identified itself as an 'experimental' programme. However, Sagynbaeva's personal vision of a hybrid Soviet–American school of sociology was gradually colonized by new plans to create a more 'Western' department, put forward by foreign academics recruited by foreign organizations to 'reform' the department. Although subsequent approaches to the experiment differed and were contested, all aimed to bypass Soviet experience and knowledge and adapt disciplinary knowledge produced in American and European societies to the local context of Kyrgyzstan. While this shift did not affect the department's innovative orientation, it segregated it from other academic institutions in the republic, distanced it from the state, and reinforced its affiliations with American culture, politics, education and funding.

AUCA is a 'private' university, though legally sanctioned by the Kyrgyz Ministry of Education, which represents itself as a model for 'independent' higher education in the republic. Its main sources of funding are tuition fees, the Open Society Institute and the US State Department. While the Kyrgyz government signed a memorandum of understanding with its main sponsors, its tacit responsibility is political non-interference. This autonomy is conditional on political relations, including those between the university and the Kyrgyz and American governments. Its fragility was exposed when an internal power struggle erupted in 2003, pitting different factions of faculty and administration against one another in a bid for institutional identity and control. The Soros Foundation, US government and Kyrgyz state all threatened to withdraw their support for the university if the issue was not resolved to their satisfaction (see Abdrakhmanova 2004).

This embeddedness in external political and economic relationships necessarily impacts on the nature and quality of academic work (Sanghera 2003). Just as sociologists working at BHU are pulled between loyalty to the state and the need to solicit supplemental funding from non-state organizations, those at AUCA are torn between commitments to 'civil society' and American higher educational institutions, and the need to garner social approval from the Kyrgyz government and society. Since its inception, therefore, the department has fought two battles on four fronts. Academics distance themselves from the Kyrgyz state and national education system while attempting to gain legitimacy within both, and align themselves with and obtain accreditation from US educational authorities while distinguishing themselves from American sociologists as members of a uniquely 'Kyrgyz' institution.

These tasks have been made particularly challenging by the more immediate problem of financial hardship. Even the comparably high salaries offered at AUCA are insufficient incentive for highly qualified instructors to stay on, and the administration does not provide the department with the resources or authority to hire new instructors despite arguments that this would enhance the university's own desire to raise its profile within the international academic community. After five years of failed appeals for financial and professional support from the university, sociologists thus began seeking alternative fundraising solutions (Ablezova 2003; AUK 1999b, 2001). One was the establishment in 2002 of the Applied Research Center, which works to broaden its support base with the foreign organizations for which it conducts research. For example, as part of a large-scale study conducted for one international organization, researchers managed to secure basic research equipment such as a laptop, camera, tape recorder and video camera (Ablezova 2003).

However, as at BHU, they often feel disempowered by the unequal relationships they have with their foreign 'clients'. One project administrator remarked,

> one organization published [our research], and we are going to publish something for another.... But we do not have any kind of rights... I mean we don't have money for it. That's why we don't ask them. For sure, we will have some credits for publishing. There should be an inscription that it was conducted by the Applied Research Center and [have] our names there.... The basic problem in our center is financial. We don't know how to negotiate these things. [There are] little things we just don't know. Because the data is not our property, it's the property of the clients.
>
> (Anonymous)

In addition, the department has increasingly turned to foreign academics for support and collaboration in curriculum development, departmental adminis- tration, sociological theory and social research methods. Having been targeted as a 'progressive institution' by Western universities and aid organizations, it hosts visiting foreign faculty on a regular basis. Although one document claims that the department intended to 'recruit well-known specialists from [the Kyrgyz Republic], the US and other western countries', new recruits have been drawn primarily from the US (AUK 1999c). This cohort, which tends to earn $200 per month or more, supplements the department's permanent Russian-speaking local faculty, most of whom receive average salaries of $80 to $150 per month.

The department also recruits through reproduction, sending promising graduates abroad to pursue further education abroad on condition that they return to Kyrgyzstan to teach. There are concerns, however, that this may contribute in the long run to 'brain drain' rather than sustainable development. Graduates with English-language degrees from Western universities, as in other postcolonial societies (Hiller 1979), often face difficult choices about whether to return to Central Asia. One American sociologist affiliated with the department argued that

> AUCA is targeting the best and the brightest to give them a ticket out.... I know you can't stop people from doing this...I fully have sympathy with the hierarchy of needs; that if you are hungry you can't think about the great philosophical issues of the universe.... I don't know if what I hear is just some sort of general academic grousing that you hear all academics doing about how tough life is in the academic world, but for them it is hard.
>
> (Anonymous)

This tension also manifests in the social organization of the department. Formally, it operates according to principles of democratic governance, mutual cooperation and academic freedom. Many of its members in fact distinguish themselves from instructors in state (also referred to as 'Soviet') universities by their deliberate refusal to reproduce traditional hierarchies of age, status and degree. Indeed, faculty meetings do often involve serious debates; younger instructors enjoy dynamic relations with older colleagues (most of whom fall into Ibraeva's 'intermediate' category of middle-aged academics and most of whom have received some degree of sociology education in American universities); team teaching is encouraged; and instructors are often invited (or, in times when teaching staff is particularly stretched, required) to design courses which are added to the curriculum without necessarily receiving formal approval from senior faculty.

However, power and status are unequally distributed within this department not according to age and academic degree, but rather by a combination of occupational status and ethnicity or citizenship. Foreign *shtatnye* [full-time] academics command considerable prestige owing to their ideological affiliation with 'the West', but exercise limited professional power as they are often poorly integrated into the university's formal and informal power structures. Their high level of job security and exclusion from indigenous power structures such as the clan, however, make them well placed to lobby the university administration on sensitive issues where their Kyrgyzstani colleagues often fear to tread. Full-time Kyrgyzstani instructors occupy a more ambiguous position, being afforded greater professional power at the university level but less intellectual and academic prestige within the department than their foreign colleagues. As their salaries are contingent upon philanthropic politics and student 'markets', they also have lower job security, which makes them reluctant to enter into disagreements with the administration. Part-time local instructors [*sovmestiteli*], particularly those who do not speak English and spend little time in the department, have the least power and prestige, and part-time, temporary foreign instructors often play very little role in department life at all. These social divisions are reinforced by the department's two-way language barrier: most foreign faculty speak little or no Russian and few local instructors speak English. Although local faculty members have taken it upon themselves to learn English, and foreigners to speak more Russian, power within the department continues to be clustered by language, with English-speaking local and foreign faculty forming an influential core.

These inequalities are reinforced by the administration's privileged treatment of foreign faculty and the absence of incentive or pressure to conform to a departmental identity. The same American academic remarked that

> especially with the young faculty, there's no senior local faculty to act as mentors that didn't live under a very autocratic system. I think [there is a] sense of faculty governance, faculty responsibility, faculty taking charge of the academic mission of the university, and at once you see that

happening, but at the same time you don't. I don't see it enough and I don't see it sustained as much as I'd like...where people, where their single-minded devotion is to making this work, and to work becomes my responsibility shared with my colleagues, a sort of collegial intensity that I think is going to be required here.

(Anonymous)

These cultural divisions have created an intellectual–technical division of labour within the department. While part-time, local and younger instructors are not systematically exploited at AUCA, they are nevertheless responsible for the vast majority of administrative and technical work. This includes writing reports for the university administration, marketing for student recruitment and liaising with other departments, as well as organizing events and translating for foreign colleagues. On the contrary, for a number of years, the intellectual content of the sociology programme was developed by full-time faculty members who made key decisions about curriculum design, course offerings and programme policy before presenting them to the department for discussion. This pattern has begun to change in recent years as local instructors take interest, gain confidence, and demand greater involvement in the production of syllabi and curriculum. However, intellectual authority remains determined by degrees of 'Western-ness', with Orientalist and Occidentalist theories of knowledge shaping collegial relationships. A similar 'foreign–local' division also therefore manifests itself in more intellectual forms, in particular, the definition of sociology itself.

Constructing sociology: teaching at AUCA

While sociologists at BHU aim to create and strengthen institutional ties between sociology and the nation-state apparatus by defining sociology as an applied profession, at AUCA the discipline has been constructed in deliberate opposition to existing conceptions of the relationship between social science and society in Kyrgyzstan. Its faculty have striven to sever associations with what Soviet defini-tions of social science are and to foster new affinities with the 'international' (here meaning Anglo-American) academic community. Sociology was therefore first defined as a liberal art oriented towards explaining and understanding society as opposed to 'fixing' it, and as a discipline that would enable students to become independent and 'critical' thinkers (AUK 1999a; Ibraeva 2003). While it is also viewed as a training ground for a new intellectual elite and fundamentally 'new type of citizen' (Sharshekeeva 2001), the meaning of this differs from the state-centric 'cadre politics' which shape conceptions of sociology at BHU. At AUCA, sociology engages not principally by applying technical skills to solving practical problems, but through producing critical research as a 'service' and social criticism.

Although the orientation towards liberal–critical sociology has been marginally dominant within the department, it is also highly contested. Neither institutional legacies nor governmental authority have played a decisive role in the making of departmental or disciplinary identity, and the diverse faculty has struggled to strike a balance that meets the expectations of both 'local' and 'foreign' constituencies. This is complicated by an underlying tension between 'internationalizing' and 'nationalizing' forces, manifested in the desire to create a Western-style sociology that will also be recognized as a legitimate, specialized profession in Kyrgyzstan. While the overarching identity as a private, post-Soviet department has provided space for the development of alternative understandings of sociology, the need to integrate the discipline into new cultural and political frameworks has also fostered increasing attention to the applied-professional model.

The early mission of the department was fluid and nebulous. Indeed, instead of being subject to strict governmental oversight, academics felt that while they were expected to produce and implement full curricula, 'there was absolutely no foundation for creating programmes at AUK . . . no goals or concepts, in general, of what [or] who we would graduate. There was absolutely nothing' (Sagynbaeva 2003). The freedom to design new parameters for sociology education outside the national standards opened spaces for intellectual and pedagogical innovation. However, in conjunction with a shortage of books, journals and communication with other sociologists, it also created an intellectual vacuum. Initial plans to develop a locally relevant intellectual vocation were thus overshadowed by the pragmatic demands of institutionalizing it as a standardized, degree-granting academic discipline. This resulted not in the articulation of a 'new Soviet–American sociology', but rather in the urgent importation of foreign, mainly American and British, models of sociology education.

The practice of adopting foreign models of sociology education became more institutionalized as the university began to hire foreign sociologists. In the autumn of 1998, for example, I was assigned to serve as Sagynbaeva's co-chair in the new department, and while we developed a good working relationship, she also felt that the department, while 'alive', never became what she initially imagined it would be. This is evident in a comparison of promotional brochures published by the department in 1999 and in 2003. The definition of sociology in the first brochure was a generic one that had been adapted from a selection of US websites, and its learning outcomes reflected the heavy influence of American sociology (AUK 1999). Sagynbaeva did not include in this brochure either her initial intention to build on Soviet sociological experience or an introduction to pressing 'social problems' as defined by other Kyrgyzstani sociologists. However, she added a list of areas in which students might find gainful employment 'developing effective solutions for complex social problems', though this too was modified from the website of the American Sociological Association. Despite her proclivity to view sociology as an intellectual vocation, Sagynbaeva's immersion in Kyrgyzstani society, concern for students' material welfare and responsibility for department finances compelled her to take a more pragmatic approach to defining the discipline.

The next brochure, published by the same faculty members in 2000, built on this image of sociology as marketable form of liberal–critical scholarship, but added to this a new form of symbolic power: identification with 'the West'. By this time, although many government officials and educators remained critical of AUK's Americanism, it had gained a reputation as an elite and internationally recognized university. However, high tuition fees, distance from traditional educational institutions, emphasis on English-language instruction and liberal education, and contentious affiliation with the Soros Foundation and US government made it difficult to attract students seeking 'marketable' professional degrees. Intellectual geopolitics hence became increasingly important in the race for enrolment and exerted greater influence on professional identity. In an attempt to distance themselves even further from the Kyrgyz state and traditional forms of Soviet education, the department emphasized its academic ties with the US and Western Europe, as well as with capitalist values more generally. The brochure asserted that sociology was recognized and practised 'in all developed countries of the world community', and that a sociology education would enable students to become mobile, independent, oriented to the 'modern world' and valued in international labour markets.

By 2002, students had successfully graduated and gone on to work and study both in the region and abroad (AUK 2001). This expansion also fostered diversification, and faculty members began to criticize earlier conceptualizations of the field. Foreign sociologists continued to promote the liberal–critical agenda, but also began to question its uncritical application to Kyrgyzstan and to develop new courses which addressed more 'national' issues or that included localized content. Many Kyrgyzstani academics, on the other hand, grew frustrated with the department's slow growth, continuing financial difficulties and societal alienation and began constructing other visions of disciplinary development. Drafts of a new promotional brochure, published in 2003 in both Russian and English, therefore concentrated less on promoting sociology's academic virtues and more on selling its practical usefulness as a tool for career advancement in what the authors defined as an increasingly competitive, 'outward-looking' society (AUK 2003b). Sociology was still represented as a way to 'understand society', but new attention was paid to how theoretical and critical insight into issues such as poverty, crime and corruption was also important for 'resolving these sorts of problems, creating theories which explain the laws of the social world, helping leaders to act and even helping to predict the future' (AUK 2003c). Finally, prospective students were assured that they would be taught by Western and Western-educated sociologists; that the curriculum had been approved by American and European specialists; and that degrees would be recognized both in Kyrgyzstan and abroad (AUK 2003c).

However, the inclusion of social problems such as corruption, the use of concepts like 'social law' and prediction, and the new emphasis on sociological professions in government, commerce and industry reflect a subtle indigenization of the field. By 2003, sociology was defined as a scholarly, practical and marketable discipline that was oriented towards public service at both national and international levels. It was legitimized not only by its grounding in liberal

114

traditions of 'critical thinking', but also by its technical practicality, national relevance and recognition from Western 'experts'. While it was initially conceptualized as an intellectual experiment within and for a transitional post-Soviet society, cultural and material factors intersected to reorient the project in two ways: first, towards American models of sociology as an academic discipline, and second, towards Kyrgyzstani models of sociology as an applied profession. The trajectory of this process can be seen not only in the department's marketing materials, but also in the successive revisions of its undergraduate curriculum. The first draft, although not the hybrid Soviet–American degree Sagynbaeva had initially envisioned (AUK 1998), ambitiously aimed to 'meet both international standards of sociological training and the particular needs and interests of university students in Kyrgyzstan' (AUK 1998a). As in BHU, students were expected to complete a certain number of subject hours of instruction, corresponding to 'points', which in turn correlated to US-style 'credits'. Students were also required to choose seven elective courses during the duration of their studies.[17] At the time of its implementation, however, there were instructors to teach just three of the nineteen proposed electives. Faculty expressed concern about this early on, suggesting that a minor specialization would be more appropriate and asking, 'do we have the human resources to announce a sociology degree programme?' (AUK 1999b) In other words, the need to 'plug gaps' in the curriculum affects private as well as state universities.

The department is further compelled to develop a full programme compatible with different educational systems. On one hand, sociologists must demonstrate to the Kyrgyz Ministry of Education that students undertake study in a wide range of 'disciplines' that will prepare them for 'theoretical, applied and pedagogical work in social science institutions, industrial enterprises, organizations and commercial-entrepreneurial structures' (Ministry of Education 1999). While the university increasingly encourages the introduction of optional courses, the ministry considers electives primarily only for 'enhancing professional quality' (Ministry of Education 1999). Government evaluators were also concerned that the department's 'new approach' to instruction, a 'synthesis of pedagogical principles generally taken from Kyrgyzstan and the US' (Ministry of Education 1999) contained few courses directly related to social problems in Kyrgyzstan and that this detracted from the degree's practicality (Ministry of Education, Attestation Commission 2002).

On the other hand, accreditation from the US is necessary if the department's degrees are to be recognized by universities abroad. American reviewers have been more concerned with issues such as eliminating what they deem excessive course requirements; the balance between theory, method and thematic course content; identifying what constitutes 'core' knowledge in sociology; and discerning what types of sociology are most necessary in Kyrgyzstan. But they, too, have consistently expressed concern about the programme's universalist approach to theoretical knowledge and relative lack of attention to the sociology of Kyrgyzstani society. Various efforts have been made to correct for this bias, including the introduction of courses on the history of sociology in Kyrgyzstan, post-communist social change, nationhood and ethnicity in Central Asia, and the politics of post-Soviet transition. However, institutionalized Occidentalism overshadows efforts to indigenize the intellectual content of the programme. Sagynbaeva remains frustrated with both the silence of localized understandings of sociology and the existence of obstacles to producing it:

> I understand that all this knowledge is, you could say, western. In the Soviet system there was little knowledge, and as yet in Kyrgyzstan there is none at all. It is all western. I would like ... so that in the courses we look at both western theories and some sort of purely Kyrgyz life.... Of course, this won't be anything grand. It won't be scientific. But they have to try.[18]

> (2003)

By 2002, there 'was a degree of consensus that there was something not quite right' about the curriculum. Balihar Sanghera, a British sociologist working for CEP, initiated discussions about curricular reform among his English-speaking colleagues shortly after his arrival to AUCA. He felt that the sociology programme was organized 'bizarrely', 'chaotically', with no apparent logic to the inclusion or exclusion of courses. There was 'no structure ... to how students would progress in the lifetime of their course'. Despite general agreement that the curricular content should be changed, there was considerable disagreement about how to do so. Sanghera aspired to a curriculum based on 'best practices' from both Britain and Eastern Europe that would give students a 'broad understanding of what sociology entails'. Others wanted to include more 'marketable' courses in areas such as quantitative research methods, and were afraid that theoretical courses would not prepare students for probable careers as market researchers.

The curriculum was revised seven times before it was finally accepted by the department and sent on to the university's curriculum committee for approval. It was the artefact of both professional and epistemological negotiation, not only between individuals within the department, but also between competing philosophies of education and conceptions of the nature and social role of social scientific knowledge. Many of the faculty were engaged in a campaign to make sociology more usable for political and economic power; others maintained a more idealistic version of relevant knowledge. Sanghera (2003) advocated a non-commercial approach. In his view, the role of a university was not to skill market researchers, but to 'broaden the horizon of undergraduates in areas that we think are useful'. However, he also admits that the new curriculum also had a more pragmatic agenda: to increase student enrolment.

As in previous years, the department's success depended on its ability to recruit qualified instructors. And, as with earlier curricula, half the programme's courses could not be taught at the time of its implementation. Because instructors are not salaried and under constant threat of having their courses cancelled due to low student enrolment, many chose to offer only 'popular' courses without regard for how these fit into a broader pedagogical framework, or to offer the same courses every term in order to reduce workload.[19] This raised intellectual as well as economic concerns, as the intellectual content of the curriculum was heavily determined by economic rather than pedagogical logic. 'Academic good practices', Sanghera argued, are intertwined with the 'whole business of economic survival'.

Disciplinarity

Striking a balance of theoretical and practical work, national and indigenous subject matter and intellectual and economic demands is an enduring challenge for sociologists at AUCA. It affects not only curriculum development and teaching but also foundational conceptions of disciplinarity. In contrast to BHU where the boundaries of sociology are clearly demarcated from those of other disciplines, intellectual parameters of sociology are more fluid and contested at AUCA. This is exemplified in successive debates over whether to merge and affiliate the department with others, including Anthropology and International and Comparative Politics (ICP). Increasing stratification between 'stronger' and 'weaker' departments and disciplines inflamed controversy about whether sociology can and should be considered an independent knowledge unit, and whether the department commands enough prestige and legitimacy to survive on its own.

In 2002, the department of International and Comparative Politics (the university's most lucrative and popular department) suggested the departments run a programme of joint admission, whereby second-year students would be required to major in either sociology or political science. From the ICP perspective, affiliation with a more 'authoritative' and established discipline would attract high-quality students to sociology and ultimately raise its prestige. Sociologists rejected this interpretation, fearing that students would ultimately choose not to major in the less prestigious field; they also felt locked into a competition for resources with the ICP department for resources, recognition and power within the university (AUK 2003d; Sanghera 2003). They thus continued to reinforce the disciplinary boundaries of sociology by emphasizing its unique contribution to social scientific knowledge and asserting its more practical usefulness for society.

A similar debate ensued in 2003 when another foreign faculty member from the Anthropology Department, then Kyrgyz Ethnology, recommended the departments unite to consolidate academic strengths and student enrolments. The proposal was finally overruled by the chair of Sociology; however, rank and file academics in both departments had been critical of the plan all along. Merging the departments would have entailed 'some rationalization and job losses', particularly among the anthropologists (Sanghera 2003), as well as the demotion of one departmental chair. Many lecturers also 'felt that anthropology and sociology were sufficiently different from one another to merit having different departments' (Sanghera 2003). As one anthropologist argued, neither discipline was

RE-DISCIPLINING KNOWLEDGE IN KYRGYZSTAN

established or well understood. Combining the programmes would therefore only lead to more confusion and the eradication of both; to *kasha* [here meaning a confused 'mess'] (also Sanghera 2003). In this interpretation, it was imperative to delineate and institutionalize them as discrete disciplines, each with its own specialized body of knowledge, skills and uses, and its own funding constituencies. The institutionalization of disciplinarity over interdisciplinarity in sociology is a deliberate (if not always reflexive or fully articulated) choice, influenced by material as well as intellectual considerations. Here, boundary-work is a matter of both professional survival and intellectual clarification, with one in fact conditioning the other.

Research

The conjunction of structural conditions and ideas has also affected the nature and organization of research conducted within the department. While AUCA began as a teaching college, faculty have been formally encouraged to conduct academic research as part of their professional responsibilities (AUK 1999a, 2003a). In 2002, the university allocated the department a small second office and appointed Mehrigiul Ablezova, a sociologist who had worked as an inter-viewer for a local social and market research company, as head of research. 'When I came here', she recalls, 'they just decided to offer me a position as the head of the sociological laboratory. And I became one. The first thing I did was rename it the Applied Research Center' (Ablezova 2003). Renaming the labora-tory was a meaningful symbolic gesture for Ablezova, who, as a member of sociology's younger generation, is sceptical of 'Soviet' and 'national' academic traditions and more comfortable with the concepts, practices and language of the American sociology in which she has been trained. Whereas the term 'laboratory' conjures images of experimental research and 'hard science', the notion of an applied research centre suggests space for the study of social problems and their practical alleviation.

Ablezova was faced with the formidable task of creating a respectable centre for research in an empty room without financial resources, academic materials, equipment or staff; indeed, there were no established research interests, and no obvious constituencies that need development. She and other faculty members therefore sought additional sources of support beyond both the university and the state, specifically from international organizations. The Applied Research Center receives no state or university funding, instead attracting research money by commissioning studies from international organizations. Its members are ambivalent about state-centred concepts of development. On the one hand, they place more faith in international agencies, which they believe are equally as dominant in the research process but less likely to manipulate research results for political purposes (Ablezova 2003). In addition, Ablezova argues that international organizations offer the centre 'moral' support because, unlike the state, they conduct training sessions, share information, and encourage the staff to 'make a difference' with their research. However, this belief stems from two things: a history of disappointment with governmental cooperation, and a general lack of trust in the state. On the other hand, therefore, researchers lament the integration of research into postcolonial development practices that often preclude the articulation of productive indigenous alliances. One, who points out that she has never worked for the government, admitted feeling that some studies she conducted were

> a waste of time, a waste of money, and a waste of talent, because I didn't see any kind of results from these projects. Nothing happened, nothing changed after that. I think that sociologists should work more actively with the government so that they can change something. For example, if we're studying poverty, we should work with, I don't know, the Ministry of Education [or] the Ministry of Social Protection in order to have some kind of power to change these things. Because sometimes, my fear is that when we conduct these surveys [for international organizations], after that, nothing happens.
>
> (Anonymous)

Nevertheless, she says that she is 'more into working with non-governmental organizations. They try to do something to change things.... So I'm more interested in working with NGOs rather than government. But government is more powerful'. Such statements reflect frustration at the discrepancy between the rhetorical role of sociology (to write prescriptions for healing social problems) and what it actually is, in her mind, the production of research that has no practical effect because the researchers do not work in cooperation with policy makers. A colleague is also disappointed in the lack of state support for sociological research:

> The state invests very little in research. Very little.... And thus on the whole a lot of research is done by international organizations that order it. Even the research on children in poverty is a commission.... Child poverty in Kyrgyzstan, and who commissions it? Foreigners. And many other projects, too. And then [the organizations] have the right over everything in this project: over all the data – we don't have the rights to one bit of information. So it turns out that there are several research projects on a single theme, because even if they publish it, they do it in the west and nothing stays in Kyrgyzstan – no publications, no nothing. It's very bad.
>
> (Anonymous)

In contrast to BHU, sociological research at AUCA is highly decentralized, even atomised to the point that, as one quipped, 'someone does something but no one knows that someone is doing something or what anyone is doing'. While there is

119

no formal 'scientific theme' for research, members of the department have developed areas of methodological expertise.[20] However, the thematic content of research is still determined almost entirely by external demand (Newman 2003). Many, however, feel that what this arrangement lacks in autonomy it makes up for by offering them opportunities to explore different questions and themes in ways that would otherwise be untenable. One young researcher, for example, explains that she enjoys her profession because 'it's always new, sociology – you can develop as much as you want, it's not something singular... everything is always interesting; you can open up everything... you can resolve various sorts of problems, and maybe help'. Another says the most appealing thing about her choice of career is that 'you never work on the same topic... you always discover something; you are so flexible within sociology'.

Sociologists at AUCA have the political freedom, both internally and externally, to determine the direction of their own sociological research. However, this does not necessarily translate into greater desire to conduct independent research. Excessive teaching for relatively low wages reduce the amount of time and energy instructors can spend on intellectual work, and lack of institutional sponsorship breeds dependence on commissions from external clients. There are few illusions about the complexities of this; as one senior faculty member pointed out, 'to do research through the support of one's institution is a luxury we still do not have.... Unfortunately, the financial question sometimes compels one to work on other things to the detriment of professional work'. These other things tend increasingly to be commercial projects that, while requiring methodological skill, lack comparable theoretical substance. Thus, while sociologists have managed to sustain research activities through commercial *zakaz*, this constrains the institutionalization of intellectual or professional culture within the university.

Boundary-work and contingency in sociology at AUCA

The activities and practices of sociologists at the AUCA offer a second example of the politics of knowledge in post-Soviet Central Asia. As with the BHU department, there is a close relationship between the rhetorical intellectual and institutional identities elaborated by academics and the existential conditions of postcolonial Central Asian society. At one level, sociologists recognize that decisions about how to define and practice sociology are constructions; for example, the existence of uneven power relations is no secret within the department. Efforts to democratize professional relationships and redistribute authority, however, have failed to deconstruct these hierarchies, which are embedded in both economic relations and colonial epistemologies. However, shared ideals about the importance of reducing the effects of power relations in the academic sphere have created space for sociologists from diverse backgrounds and intellectual orientations to take different positions within the department, thus enabling the emergence and development of alternative visions of the discipline: sociology as liberal–critical scholarship and sociology as applied service profession. While these remain mutually independent, movements to rethink the relationship between teaching and research and the increasing participation of Kyrgyzstani faculty in the intellectual life of the department suggest that the two approaches may converge in the future. Whether one type of sociology becomes ascendant over the other depends on a variety of factors. The relative dominance of the liberal–critical scholarship model of sociology has been maintained by the university's identification with American liberal arts education and its rejection of Soviet models of technocratic education; a continuing supply of sociologists from American and British institutions; the routine training of younger faculty members in American universities; the prevalence of liberal and critical scholarly materials in the departmental library; and external moral support for academic innovation. Changes in these conditions would present sociologists with a new set of problems and choices in constructing the boundaries and role of the field.

The establishment of scientific legitimacy for sociology here is relatively unproblematic: it is asserted on the basis of the department's affiliation with 'Western' sociological traditions and institutions which are represented as universally authoritative. The more pressing problem is communicating the discipline's social relevance, not only to prospective students, but also to the Kyrgyz state (to which the department must appear experimental but not politically or culturally threatening) and to American educational authorities (to whom the department must exhibit both its international and national 'qualifications' and its commitment to post-Soviet, specifically anti-socialist social change). In other words, sociology at AUCA must be at once a discipline dedicated to preserving and revolutionizing the social order, and sociologists must strike delicate balances between different expectations of what constitutes legitimate social scientific knowledge in both contexts.

The need to appeal simultaneously to different constituencies has forced faculty members to modify descriptions of their teaching activities and to reconcile minimum compliance with national educational standards with maximum compliance to the professional and intellectual norms of Anglo-American sociology. The AUCA department has absorbed some of this tension by creating a division of labour between teaching and research, with a curriculum oriented outward, towards 'the west', and research oriented inward, towards the study of social problems in Kyrgyzstan. However, this more nationally orientated research receives little professional recognition from other social scientists in the republic because it is maintained predominantly through commissions for foreign clients, many of which advocate non-state or 'civil society' solutions to national social problems. Thus, while the department would like to be perceived as a non-political educational establishment, its self-imposed distance from the state and state-run sociology institutions gives it a highly political, even oppositional profile. Instead of countering this, though, the department capitalizes on it by emphasizing its unique contributions to the liberal functions of sociology education, including personal development, professional growth, individualized critical thinking and social enlightenment. The nature and role of sociology in this context is constructed in direct relation to the department's reformist vision of social change.

Sociology at AUCA has never been 'in transition' as this phrase is often interpreted in post-Soviet institutions. It was imagined and introduced as innovative in Central Asia; a field whose philosophy and geopolitical affiliations enabled it to reject delegitimized relations and epistemologies and embrace the new. From its inception, it self-consciously represented an ideal rather than a change. However, historically familiar patterns of academic dependency have emerged within the department; neither the type of sociology being institutionalized here nor the structural and cultural contingencies of the process are entirely new. Sociology at AUCA is in a different type of transition as faculty members struggle to negotiate ambiguous balances between East and West, national and international, theoretical and applied sociology, and intellectual and financial autonomy and dependence. As such, it embodies the problems faced by those attempting to institutionalize a liberal–critical scholarship model of sociology in a technocratic and aid-dependent postcolonial state.

7

PUBLIC SOCIAL SCIENCE IN CENTRAL ASIA

The case studies presented in Chapter 6 illustrate how alternative conceptions of academic sociology have emerged within different institutional contexts in post-Soviet Kyrgyzstan. They also demonstrate that 'Central Asian sociology', far from being a monolithic enterprise, in fact encompasses a range of meanings. However, the field also extends beyond academic institutions into the public sphere, particularly into areas of commerce, 'development' and media. This chapter will explore how and why the content, methods and roles of social scientific knowledge have been negotiated in the national print media, thus becoming types of 'public science' and tied to debates about independence, decolonization, authoritarianism, globalization, Westernization, modernity and truth.

Sociology in the national press

The association of social scientific knowledge with non-academic institutions was a long-standing Soviet practice. However, the deliberate use of sociological debates as public platforms for professional and intellectual politics is a more recent phenomenon. It emerges at the intersection of a number of factors: the expansion of independent media outlets and reduction of scholarly forums, increased public concern about knowing social 'truth', the reorganization of science and higher education, the ideological association of organized knowledge with modernity and development, competition between different sociological groups and institutions, and sociologists' quest for legitimacy and support from diverse constituencies, including the state, international organizations, business and public citizens. The immediate socio-political contexts for this discourse were the post-Soviet privatization of land and other state property, and the formal (albeit far from substantive) transformation of centralized, authoritarian politics into a democratic system in which power is distributed equally between citizens and elites. The practice of defining, popularizing and defending sociology in the media must be understood in terms of how discourses on method and ethos, as well as the more general relationship between science and politics, were constructed within these broader contexts.

The term 'public sociology', particularly in American lexicon, most commonly refers to an engagement between academic sociology and public, often social or political, commitments; it can also refer to efforts to bridge the gap between academic scholarship and public consciousness and action (Burawoy 2005). However, it can also refer more analytically to the rhetoric used to popularize, legitimize and garner support for science itself. In contemporary Kyrgyzstan, it means both. Sociologists use popular media as sites to 'construct ideologies [of social science] with style and content well suited to the advancement or protection of their professional authority' (Gieryn 1983: 783) while simultaneously attempting to educate the public about sociology and make political statements about current events.[1]

The division between academic and public sociology has become particularly contested in the post-independence period; it is still being articulated in practice. Sociologists have ambivalent feelings about publishing in newspapers. Many see such writing as inferior to that published in academic journals or non-peer reviewed, institutionally produced conference proceedings. However, the absence of peer-reviewed journals and lack of resources for publishing in general have created a situation in which newspaper publications are often classified as 'scientific publications'.[2]

Articles about sociology began to appear regularly on the pages of national newspapers such as *Slovo Kyrgyzstana* (a pro-government publication), *ResPublica* (an opposition paper) and *Svobodnye gory* (the newspaper of the *Jogorku Kenesh*, or parliament) shortly before independence.[3] While all part of the larger rise of public social science, two debates in particular stand out as significant. The first, a series of articles about public opinion on privatization in Kyrgyzstan, demonstrates how the tension between scientific objectivity and political interest in social scientific work has been reconstrued in the post-Soviet period. It revolved around sociologists' authority, or lack thereof, to analyse, evaluate and criticize controversial government policies, and illustrates how representations of the relationship between sociology and the state were realigned as the latter became more authoritarian (Spector 2004). It also reveals that two types of boundary-work (to expand scientific authority and to protect the autonomy of scientific knowledge from the political field) were seen as vital by sociologists.

The second major debate, which became known as the 'ratings scandal', consists of articles about 'political ratings' published by Isaev and associated researchers from 1993 to 1997, as well as critiques written by sociologists from other institutions in the republic. This debate embodies a range of topics related to the problem of scientific credibility as it is manifested both among and between sociologists, the broader public and the national power elite. It reinforces that the boundary between science and politics is often negotiable in the face of increasing political pressures on sociologists, as well as how rhetorical dimensions of scientificity such as 'objectivity' may be defined strategically in relation to political and professional circumstances.

Such public debates about academic knowledge reflect deeper controversies about the role of the field, its epistemological foundations, and its emerging professional ethos. Because they centre on studies which take the methodological form of public opinion research, before turning to them it is important to understand how the study of public opinion has been defined and practised in Kyrgyzstan and what relationship it bears to sociology more generally.

Public opinion: the 'democratic' face of sociology

The vast majority of sociological research conducted in Central Asia during the 1990s consisted of survey research in one form or another. Just before Uzbekistan' independence, for example, the director of the Tashkent Office of Public Opinion under the SSA argued that the rapid proliferation of survey research in the region during perestroika had raised as many questions as it had answered, including, 'What is this science, the sociology of public opinion? How do sociologists get to and analyze the data? What are the practical applications of results and their analysis?' (Chernyshev in Luk'ianova 1990: 55). Such questions about the definition of legitimate knowledge and the authority of legitimate knowers had also become highly contentious in Russia, Kazakhstan and Kyrgyzstan by the mid-1990s.

Despite such fundamental uncertainties, opinion research rose to ascendance in sociology. In fact, the term 'public opinion research' refers to diverse methodological practices, including polls, rating and ranking surveys, structured interviews and focus groups. The common denominator is that each of these methods may be used to ascertain the 'public mood' about social issues. A number of factors contribute to the popularity of this approach: Marxist–Leninist philosophies of social science as a technology for scientific management, theories of knowledge that empirical data constitute a scientific antidote to the political manipulation of social reality, methodological individualism, and the conviction that public opinion research is an integral element of modern, democratic and civil societies.

The concept of public opinion, however, is not merely historical. Owing to its focus on the 'subjective individual' as opposed to structural social forces, during perestroika it was represented as the 'democratic' face of Soviet sociology. Its origins are rooted in an earlier Communist Party practice of conducting surveys to provide party leaders with 'feedback' about how Soviet citizens 'understood' top-down political and economic decisions and to ascertain whether they were 'ready' for certain types of social reforms. The most famous of these were short questionnaires conducted by the newspaper *Komsomol'skaia Pravda* in the 1950s and 1960s, which readers could voluntarily clip out, complete and return to the editors (Buckley 1999: 224). Data from such non-representative surveys were used to ascertain and, it was argued, influence the 'collective social consciousness'. They were intended, among other things, to legitimize the asymmetrical relationship between the Communist Party leadership and society at large, in which 'the masses would learn the truth about society from the party through its propaganda, and the party would learn where and when people would be prepared for social change, as well as new techniques and strategies for "revolutionary" struggle' (Inkeles 1958: 18).

Censorship on the critical analysis of 'sensitive' questions meant that early studies of public opinion in Soviet Kirgizia were neither systematic nor statistical and lacked methodological rigour (Buckley 1999). 'In past years', recalled one practitioner, 'we held the view that public opinion meant letters and announcements directed by citizens to party and Soviet organs. Many letters – that was good. It meant that we studied and knew [about] public opinion' (Luk'ianova 1990: 55). In other words, it was not an analytical question, but a matter of studying reactions 'primarily to determine the pace and speed of [one's] own actions. The goal [was] not to cater to public opinion, but to move it along with you as rapidly as possible without undermining your popular support' (Inkeles 1958: 24). A similar instrumental rationality can also be seen in many studies of industrial sociology during this period, particularly as industrial sociologists made heavy use of survey research for much the same purpose.

During perestroika in Kirgizia, public opinion research was reconceived as a populist counterweight to ideological hegemony in the political arena and presented as a way for people (both within and outside the party) to empower themselves with information and disempower government authorities who continued to monopolize representations of truth about social reality (Luk'ianova 1990). However, 'public opinion', defined as the sum of individual opinions of private citizens (Isaev et al. 1997) and the conversion of these into a collective consciousness (Isaev 1995) came to be perceived as a legitimate political force *sui generis* as well as a reflection of social experience. It was viewed as a social as well as a scientific institution (Bekturganov 1994: 15; Lokteva 1991) and a 'mirror in which most people's relation to power . . . is reflected' (Sydykova 1998; see also Isaev et al. 1996b).

This new, more critical function of public opinion research was first articulated as a way to democratize the Communist Party. Leninist rhetoric about the importance of information in democratic centralism and the power of mass participation were revived, and the public were reminded that 'it is well known that V. I. Lenin repeatedly said that the party leadership must have before them a full picture of the work of local organizations, as without information it is impossible to centralize party leadership' (Bekturganov 1990: 107). If party authorities did not consider the ideas of subordinates within the organization, it was asked, how did they intend to democratize their relationship with society at large?

This narrow application, however, soon broadened to incorporate other social institutions, and public opinion became prominent in discourses about the democratization of Soviet society more generally, including in Central Asia. Its reformed role was ambitious: it was argued that 'the political significance of research in public opinion is linked above all to the necessity of democratizing and humanizing socialist society' (Bekturganov 1990: 107), and that the failures of perestroika could be attributed in large part to a lack of organized knowledge about everyday life (Bekturganov 1994; Isaev and Bekturganov 1990: 3).

If public opinion was regarded as a new scientific and political force in late-socialist Kirgizstani society, then professionalizing, institutionalizing and publicizing public opinion research was the new mission of sociologists.

126

Prominent academics advocated a greater role for research, and consequently for themselves, in social and political life. Instead of being mere surveyors of general attitudes, it was argued that 'sociological groups and bureaus established within the Councils of Peoples' Deputies, party committees and social organizations should become integral parts of the effective activities of these same organizations' (Bekturganov 1990: 110). During perestroika, public opinion research therefore symbolically redirected the locus of authority away from the Communist Party and back towards 'the people'. In order to legitimize this, Marxist–Leninist theories of scientific development were invoked to assert that objective social forces are reflected in mass psychology and could be created or reformed by enlightened intervention. Sociologists advocated the creation of new alliances between state, party and academics, teams to work in a 'range of organizational offices for the study of public opinion in the regions, connected with state, party and social organizations' and staffed by a new *aktiv* [cohort of party activists] of *anketery* [surveyors] who would be selected by the party and trained by the republic's few professional sociologists (Bekturganov 1990: 110).

While several efforts were made to institutionalize this movement, as with industrial sociology, the study of public opinion was never institutionalized as a sustainable academic practice in Soviet Kirgizia, and in 1990, its most public practitioner made a devastating critique of the state of the field (Bekturganov 1990: 106). However, independence saw a dramatic rise in the number of public opinion surveys being conducted in the republic (Toktosunova and Sukhanova 1990; see Buckley 1999 for similar trends in the RSFSR).[4] Post-Soviet rhetoric about the relevance of public opinion was reinforced by the powerful symbolic association of opinion research with democratic – and by implication anti-Soviet – politics. It was asserted that 'in truly democratic countries, politicians pay attention to the results of public opinion so they are prompted to action in deciding internal and external politics of the state' (Isaev *et al.* 1996b) and that 'in civilized countries, public opinion is a political institution that is a recognized and legalized mechanism at all levels of the decision making process' (Lokteva 1991). While such proclamations are rarely supported with references to empirical evidence, they are also rarely if ever challenged. For many, the very existence of public opinion research is a clear indication of modernization and development.

Some are critical of the phenomenon. Isaev, for example, has pointed to a discrepancy between the proliferation of public opinion surveys and general confusion about their purpose, linking this with the underdevelopment of sociology more generally (Isaev, Akmatova and Dosalieva 1996). According to Bekturganov and Tishin, the lack of a creative indigenous sociological theory, the paucity of social scientific language to describe social phenomena, the lack of trained specialists in sociology, personal ambition, and the dominance of 'percent-o-mania' and '*anket*-o-mania' at the expense of more 'serious' mathematical and statistical forms of data analysis have led to sociological 'illiteracy' within the community and to subjective, and therefore invalid, research on public opinion. This in turn has negative consequences not only for the discipline's public image but also for the possibility that sociologists will be recruited as professional consultants or 'experts' in public life (Bekturganov *et al.* 1994). In a critique of Isaev's own research, Bekturganov (1994a) waged a more serious criticism: that methodological weakness, along with the conflation of sociological research and political interest were 'distorting the principles of correctly selecting the experts, methods, techniques and procedures defining political ratings'.

Veteran sociologists have been particularly critical of the ascendance of what they call 'dilettantism' and its deleterious impacts on the status and legitimacy of academic sociology in the republic. On the one hand, they argue that the expansion of sociological discourse in the national media was a consequence of its democratization. In this sense, they acknowledge that the long struggle to lift the censorship of sociology and social criticism had to some extent come to fruition after independence. On the other hand, however, they express concern that this more 'public' identity blurs the boundaries between scientific and non-scientific work and professional and lay knowledge (Bekturganov *et al.* 1994). As Tishin (1998: 32) remarked in a critique directed specifically at Isaev's work,

> sociological dilettantism emerged on the wave of high-quality sociological research and exists to this day, discrediting sociology.... Sociological material in the republic was held back by ignorance and was very rarely printed.... In 1993–4, the other extreme developed. Monthly sociological columns and weekly sociological reviews with puzzling rubrics appeared in the periodical press. The philistine style of the materials, the lack of content, and the advertising-like presentation have created an impression of political prostitution on the part of individual sociological researchers.... However, not all sociologists have fallen into this trap; [for example], the National Academy of Science has conducted fundamental sociological research about the development of international relations and the problems of national conflict and tension.

Here, Tishin clearly distinguishes between legitimate social science done within a previously hegemonic institution and 'amateur' or 'pseudo' research done by individuals lacking proper academic training or institutional affiliation, a competitive group created through the deregulation of academic activity and by the ascendance of populist conceptions of sociology that equate sociological research with the distribution and evaluation of questionnaires. One who herself worked for several years as a marketing researcher before joining a university department affirmed this critique:

> now...many people who have nothing to do with marketing sociology conduct surveys, conduct research, without even knowing how to do sampling or design a questionnaire, and what's the rule in the field, how they should conduct interviews, how they should analyze data.... And I think that it also has a negative influence on sociology.
>
> (Anonymous)

The distinction between 'scientific' and 'unscientific' analysis, and more narrowly between legitimate and pseudo-sociology, became particularly important as a disparate variety of intellectual and political actors, many of whom competing for similar positions of social power, claimed to be validated by the authority of science. There is, in addition, a generalized and almost conspiratorial fear that the results of public opinion research might be misused to enable the psychological manipulation of society at large (Isaev 1998; Sydykova 1998). Underlying this is a philosophy of knowledge which assumes there is such an entity as 'public opin- ion' and that it actually constitutes an objective and potentially powerful political force, thus making concerns about professionalism and distinctions between 'real' and 'pseudo' science of heightened significance in the republic (Bakir Uluu 1997; Bekturganov 1994a; Isaev et al. 1994b, 1997a). The importance of this issue is revealed explicitly in Kyrgyzstani sociologists' rhetorical efforts to sepa- rate legitimate sociology from pseudoscience and distinguish between 'good' and 'bad' sociological practice in the media.

From privatization to *prikhvatizatsiia*: sociology confronts the state[5]

As illustrated in previous chapters, the definitions of sociological method and practice in Central Asia have been historically contingent and often pragmatically defined. This case, based on a series of newspaper articles about public opinion research on post-Soviet privatization in Kyrgyzstan, explores how sociologists have constructed boundaries between science and politics in order to enhance the field's legitimacy within shifting political conditions.[6] Unlike cases in which aca- demics engage in rhetorical debates with one another to gain access to social and material resources, the boundary-work in these articles was targeted at counter- acting political ideologies and the spectre of authoritarianism, as well as disentangling sociology from both.[7] The challenge to sociologists' authority in this case comes not from other academic disciplines or social practices, but from the possibility of a deepening of heteronomy between social scientific knowledge and new forms of illegitimate political and economic power.

There is very little sociological research on post-Soviet privatization in Kyrgyzstan. While economists and legal scholars have expressed some interest in the topic, primarily in terms of evaluating the effectiveness of economic liberalization pro- grammes (e.g. Dabrowski et al. 1995; Nichols 1997), sociologists and anthropologists have generally neglected it. Much existing research reproduces a now-waning ideological consensus that Kyrgyzstan's relative 'success' in privatizing state-owned property, along with its 'pro-Western' and 'democratizing' government, made the republic an 'oasis of democracy and social peace in a region wrecked by powerful ethnic and religious conflicts' (Dabrowski et al. 1995: 269). The process of privatiza- tion in Kyrgyzstan is therefore often defined by its formal components and analysed according to fiscal outcomes, and the social and personal effects of privatization programmes on the everyday lives of ordinary people has often been overlooked.[8]

These experiences, however, were not lost on Kyrgyzstani sociologists, some of which had begun to investigate the social face of privatization in the early 1990s. Members of Isaev's laboratory, who had by this time begun to call themselves the 'Independent Group of Sociologists', conducted a series of surveys of public opinion about privatization which were intended to 'gather a wealth of material for rethinking, administration, and decision-making' (Isaev 1994).[9] Some results were published in newspapers simply as tables or statistical indicators (e.g. the percentage of respondents who felt privatization was beneficial or detrimental, classified according to ethnic group, class or age). In other cases, however, statistical data merely served as points of departure for political, often polemical critiques about wider concerns, such as inequalities in the privatization process, corruption and social misinformation. This space was also used for discursive struggles over the meaning of sociological knowledge itself.

For example, the introduction to the first article in the series, entitled 'Privatization for what and for whom?' was a political treatise on the social relevance of sociological knowledge as much as a description of the research results. After arguing that successful privatization depended on widespread public participation, Isaev criticized authorities for failing to 'take this into account' and asserted that sociological research was the best – indeed the only – way to obtain information about how privatization was actually progressing (Isaev 1994). The success of economic policy was still seen as being heavily dependent on the construction of ideological consensus, and the possibility of creating the latter depended on obtaining data from effective surveys of public opinion. On this basis, it was argued that 'applied sociology, if its results are used intelligently...can become an accurate barometer, precisely indicating the ways and means of constructing market relations' (Isaev 1994).

These studies were originally sponsored by sources identified only as *nomenklatura* [wealthy patrons with political power] and in one case by 'an American agency' (Isaev 1996a, 1998). However, according to Isaev (1998) 'when they realised that public opinion was shifting from privatization to *prikhvatizatsiia*, the *nomenklatura* eventually withdrew [funding]'. As surveys on privatization continued, sociologists became more critical of both the process and those implementing the changes. This eventually led to a struggle between sociologists and politicians for control over how the motivations for and consequences of privatization were defined. Early articles focused on the ineffectiveness of the Kyrgyz government's 'propaganda campaign' to popularize voucher-style initiatives (Isaev 1994a), the 'class dynamics of public opinion' about privatization (Isaev and Abylgazieva 1994) and the 'complexities and contradictions' that seemed inherent in the privatization process (Isaev, Akbagynova and Abylagazieva 1994). They also began to report percentages of people purportedly living below the poverty line and were highly critical of the overall outcomes of privatization initiatives. Gradually, the articles suggested that there were substantial discrepancies between people's expectations of progressive change and the actual results of privatization (Isaev 1994b). They began to emphasize ethnic differences in this

130

experience,[10] and bypass the analysis of what people thought about privatization to comment on why they were so passive in the movement. Several pointed out that 'the majority of people care about basic living problems and not privatization', an obvious reference to extreme levels of poverty in the republic (Isaev and Abylgazieva 1994; Isaev, Akbagynova and Abylagazieva 1994).

These arguments were more than social science, also being critiques of 'imported' policies. The sixth article blatantly asserted that 'propaganda and agitation won't work on social consciousness. It is thought that the main reason for people's passivity in privatization is serious opposition to the socio-economic mechanisms of the transition to a market economy' (Osmonalieva 1994).

By summer 1994, sociologists had published seven progressively critical articles about privatization in the republic. While the articles had 'scientific' status because they were written by academics, they were also deliberately political documents, formulated in an emotive rhetoric that combined academic jargon, social research and political platforms and analysis. Ultimately, they were deemed politically threatening and sponsorship for large-scale public opinion research studies at BHU was withdrawn in the second half of the year (Sydykova 1998). While not specifically intended to contribute to the realignment of knowledge and power in post-Soviet Kyrgyzstan, the boundary-work conducted through this experience was nevertheless influential in this process.

When the researchers began their studies on privatization, they portrayed sociology as legitimate and relevant, but increasingly found themselves doing so in a state that, while no longer Soviet, remained hostile to ideological challenges from the social sciences, and in an economy in which material resources for research and teaching were scarce and in high demand. They adapted by portraying their work as objective, scientific, politically potent and methodologically anti-political, erecting unambiguous boundaries between sociology and the illegitimately political, and asserting their relevance for the creation of a new type of scientific politics. Sociologists adopted the formal rhetoric of democratization to wage subtle critiques of the government's trends towards authoritarianism, in particular its deleterious effect on sociology, while maintaining that sociology could contribute to alleviating abuses of power by being closely allied with the state. The characterization of sociological research outlined in the articles on privatization – large-scale, empirical, methodologically rigorous, longitudinal and by implication expensive – was also linked to renewed demands for the creation of academic institutions in which such research could be conducted, and for the establishment of 'democratic' social, political and economic institutions which could become consumers for its products (Isaev 1994).

In this milieu, sociologists simultaneously produced two rhetorical discourses on knowledge. On the one hand, the articles promoted images of sociology that aimed to extend and expand the influence of sociology into two arenas historically dominated by state and party: economic policy and 'reality management'. This was reflected, for example, in statements that data were intended both to inform public opinion about privatization and to provide decision makers with information about public perceptions of the policies. In this discourse, researchers clearly identified themselves with the administrative apparatus of the Kyrgyz state. 'The results', they claimed, 'will offer scientific–informative help to power structures of the Kyrgyz state in elaborating policies for social transformation, corresponding with the socio-cultural characteristics of our republic and the particularities of the mass consciousness of its citizens' (Isaev 1994). Here, the relationship between knowledge and power is carefully constructed so that 'scientific' knowledge can be legitimized as long as it contributes to the establishment and maintenance of 'just authority' in the form of democratic governance and state-sanctioned social planning. As sociologists continued to seek patronage from the post-Soviet state, they adopted its official ideology of democratization and liberalization and framed social criticisms within this sanctioned rhetorical framework.

On the other hand, however, the researchers also aimed to distinguish sociology from politics and political activities, attempting to establish the autonomy of sociological knowledge from 'non-scientific' forms of information and argumentation. Sociological knowledge was carefully characterized as meeting four major criteria of 'democratic' knowledge: (1) 'complete, systematic and complex', (2) 'authentic, scientific and methodologically grounded', (3) 'efficient and regularly replenished' and (4) 'able to apply different approaches to studying different regions of the republic' (Isaev 1994). Appeals to the logic of scientific objectivity and neutrality became part of attempts not only to differentiate social research about political issues from political activity, but also to carve out spaces for social critique in a period when public criticism was being increasingly suppressed. While there is nothing inherently 'scientific' about many of the arguments made on the basis of public opinion surveys on privatization, connecting these claims to the methods of scientific research, and symbolically to cultural values of democracy and truth, enabled sociologists to claim a degree of intellectual authority and political immunity.

In the case of the privatization surveys, however, these discourses had little tangible impact on the personal, political and material interests of the organizations funding the research. This exercise in public science and boundary-work neither engendered changes in the actual relationship between sociologists and the power elite nor prevented powerful sponsors from withdrawing their patronage when the results and interpretations of the research ceased to validate their own agenda.

Political ratings in Kyrgyzstan: real sociology and 'pseudosociology'

Another series of articles about the 'sociology of the elite', based on surveys conducted to ascertain the relative popularity of politicians and political parties, became the site of a different type of struggle over scientific authority and legitimacy within the social science community itself.[11] This case provides another illustration of how the nature and role of sociology is constructed differently according to shifts in social demand, and how the definition of theoretical and normative concepts such as 'objectivity', 'value-neutrality' and 'good' and 'bad' sociology are fluid and emerging.

The study of 'political ratings' was popularized in Kyrgyzstan during perestroika as power devolved from central to locally elected authorities (Isaev 1991b) and local elections raised questions about predicting and monitoring political action. Ratings of political leaders in Kyrgyzstan, generated by asking respondents to rank candidates on a five-point scale, were published regularly from the early 1990s and continue in modified form to the present day.[12] While members of the BHU laboratory published more than twenty articles on this theme in a four-year period, researchers in other institutions also engaged in debates about the purpose, methodology and interpretation of surveys conducted to establish political ratings.

Isaev and others devoted considerable attention in these articles to promoting the social relevance of sociology in general and studies of political ratings in particular. The latter are justified symbolically, often with reference to their usage in 'the West'. Sociologists argue that such research constitutes an integral part of modern democratic states in which political life 'is strongly influenced by the personal quality of its leaders' (Group 1993; see also Isaev 1999c) and where 'research results serve as a believable source of social information for making decisions or correcting the political behaviour of leaders' (Isaev et al. 1994b). It is also associated with the rationalization of the political process, as ratings surveys are designed to measure political prestige and legitimacy 'not according to...position in the hierarchy, but according to...concrete deeds and the effectiveness of work' undertaken (Isaev, Niyazov and Zhigitekov 1994). It is believed that empirically derived information is not only a corrective for poorly conceived policies that fail to address actual social problems (Isaev et al. 1994b), but also a way to provide the public with vital political information 'when the parties don't and leaders won't' (Isaev et al. 1994d).

As with the studies on privatization, the realization that the administration was becoming more rather than less authoritarian shaped the rhetoric that sociologists used to represent their work in the public sphere. Justifications for the social relevance of sociology shifted in the mid-1990s from emphasizing the therapeutic and policy oriented role of sociology in modern democratic societies to focusing on its more critical functions in authoritarian states (Isaev et al. 1994b,e). While maintaining that the association between sociology, modernity, democratic politics and 'civilization' was an ideal to aspire to, academics also argued that in times of political repression, sociological research offers an anti-political challenge the hegemonic ideologies of those in power. In a 'non-objective' (i.e. politicized) world, sociologists defined their responsibility as 'analyzing and commenting on facts, not reconstructing reality, and not refuelling elements of "lies" of political consciousness, not creating illusions tied to politics' (Isaev, Ibraeva and Madaliev 1995). By the late 1990s, notions that sociology should contribute to the efficacy of state power had been supplemented with the assertion that it was necessary so as not to leave 'the sphere of the production of political products only to individual politicians, and in order to escape from the systematic and even outright manipulation of certain points of view' (Isaev *et al.* 1997a).

As with the privatization studies, and as elsewhere in the former Soviet Union at the time (Butenko 2002), sociologists used the articles on political ratings to discursively articulate a new relationship to the state during this period. Sociological research was even ascribed national missionary status, crystallizing academics' new role as an alternative power base in Kyrgyzstani society:

> The results of our research may not 'suit' someone and might be 'uncomfortable', but without servility or care for authority, without consideration of the preferred market, we will absolutely inform them about the wider community. We see this as our mission – you know we answer to national socio-political science. This is the civic and scientific position we intend to stand by, regardless of all shades of opponents and individuals.
>
> (Isaev, Ibraeva and Madaliev 1995)

In tying together intellectual, moral and professional responsibilities, sociologists also constructed new professional identities as national heroes struggling to honour the scientific pursuit of truth in an atmosphere of ideological manipulation:

> From the time we began empirical sociological ratings of political workers (since 1991), various toadies, lackies, advisors, intriguers and envious people... have tried to ruin this research from the start. But we continue with our sociological scientific studies.... In a situation where the systemic crisis of society is deepening, the social status of the population is worsening, and faith in the power structures is decreasing, the task of defining the ratings of political workers demands courage from researchers.
>
> (Isaev *et al.* 1997a)

Hence, while studies of political popularity were justified within the BHU Sociology Department as part of its larger research project oriented towards advising the nation's power elite (see Chapter 6), they were more publicly justified as a challenge to the power of this very group.

Underlying epistemologies

In this debate, arguments about the social relevance of sociology are based on a number of epistemological assumptions about the nature of scientific knowledge, its relationship to modernization, and the proper interaction of social science and society. First, social science is portrayed both as a symbolic indicator and as a practical method for the modernization of political life, particularly the rationalization of political behaviour and the transition from personality based politics to rational-action or deeds-based politics. Second, the practice of constructing social scientific knowledge is defined as objective. Valid knowledge is believed to transcend political and intellectual contingencies within the society and to act as a guarantor of truths about social reality, which are exploited by political actors that value truth for its use value and not as an end in itself. Third, this transcendental epistemology requires that sociologists be seen speaking truth to power. In a hypothetically democratic society that respects the value of truth and its role in effective social policy, they should be immune to political retribution. Finally, the production of sociological knowledge is not tied to any particular political or social system. While it is described as an integral part of democratic societies that ostensibly base political decisions on scientific research, it is also seen as a necessary presence in non-democratic societies as an alternative to the ideological hegemony of authoritarian regimes.

However, there are tensions within these assumptions which become visible in public contests for scientific legitimacy in Kyrgyzstan. In the articles on political ratings, this struggle is manifested in three separate debates: about the methodology of sociological research, the relationship between science and politics, and the reconstruction of professional ethos.

Debates over social research methods, including approaches to sampling, questionnaire construction, the interpretation of raw statistical data and the use of 'expert' or 'mass' surveys assume a prominent place in public science in Kyrgyzstan. They are related not only to questions about the reliability and validity of truth-claims, but also to issues of intellectual integrity, conformity to ambiguously defined professional norms, and the politicization of social scientific knowledge. Boundary-work is therefore conducted to create professional norms for sociologists and to strengthen relationships of trust between sociologists and the public.

From the outset, publications on political ratings included lengthy, albeit selective, explanations of the methodology that was used to obtain the results. This is partly a form of 'political education', a Soviet practice of using newspapers, radio and television programmes as media for ideological 'agitation'.[13] The post-Soviet practice of using the media as a tool for educating and persuading the public to value the production and application of survey research on political ratings resembles this older practice. Indeed, this is reflected in authors' claims that the articles 'fulfil enlightening, socialization and mobilizing functions' and that 'as benevolent bearers of social scientific information, they do significantly more to explain the essence of reforms being carried out in the state than do the acts of informers, time-servers or cowards' (Isaev *et al.* 1997a). Because sociologists consider themselves public figures, many take seriously the responsibility of public education, albeit at an acceptable distance and in the role of 'experts' (Isaev 1993a).

There are, however, other factors motivating the publication of sociological methods. Chief among them is the symbolic power of the rhetoric of transparency, which is invoked in the bitter competition for professional authority that has emerged within the small, mainly indigenous, community of sociologists in the republic. From 1993 to 1997, Isaev and colleagues published articles criticizing the research of Osh-based academic Tursunbai Bakir Uluu, branding it 'ignorant' and 'sociologically illiterate' (Isaev *et al.* 1993a). Bakir Uluu responded with articles attacking Isaev's work, labelling it 'unscientific', 'narcissistic' and 'ideological' (1994, 1997). He was joined by KNU sociologists, who suggested that the work was part of a 'political game' and that 'unqualified' research in general was a threat to the status of the discipline in the republic (Bekturganov 1994a).

Interestingly, each antagonist made similar accusations of the others. While there was a broad consensus about the imperative of certain methodological norms – in particular, that sociological research should be objective, politically detached, representative, valid and reliable – there was considerable disagreement about what each of these terms meant and where to draw the boundaries of 'correct' interpretation and professional practice. Because the representativeness of sampling techniques in social research is linked to ideals of both scientific truth and the democratization of knowledge (Blum 1991), sampling used in studies of political ratings has been particularly contentious. Flawed or suspect sampling frames are tantamount to scientific incompetence and wilful politicization, and hence, to professional betrayal (Bekturganov 1994a).

However, competing logics also shape methodological decisions. For example, while use of a five-point scale is in one instance justified on grounds of clarity for a lay readership and improving public understandings of science (Isaev 1993), in another it is defined as populist and therefore 'unscientific' by those favouring more 'scientific' statistical analysis (Bekturganov *et al.* 1994). This complexity is also visible in the different types of surveys used to ascertain opinions about leading politicians. Kyrgyzstani sociologists make clear distinctions between 'expert' and 'mass' surveys. The first is a set of questions asked of carefully selected people who 'work professionally in an area of real activity of interest to sociologists'. However, it may also mean, *inter alia*, 'people who are completely knowledgeable about politics and professionally familiar with the politicians of the republic' (Isaev, Ibraeva and Madaliev 1995), people 'chosen based on their professional background for scientific purposes' (Isaev 1998) and 'unbiased, neutral opponents who are equidistant from the powers-that-be and the opposition, who are professionals in their work, scholars who always distinguish between critical relations to power and opposition, and who have their own independent and objective opinion about the processes of social life' (Isaev 1999b). 'Mass' surveys, on the other hand, are defined as questionnaires distributed to 'simple respondents', or 'anyone living in the republic, chosen by special a method depending on their sex, age, nationality [ethnicity], education, region of residence, and other indicators' (Isaev *et al.* 1993). While the latter are generally considered more statistically accurate (Isaev and Ibraeva 1995), the former have become popular because researchers lack the financial and human resources to conduct large-scale representative surveys of the general population.

The rhetorical distinction between expert and lay knowers has been used to deflect criticism that certain studies are insufficiently representative and that research data are insignificant or inconsistent (Abdyrashev 1994; Bakir Uluu 1997; Bekturganov 1994a). In some cases, differences between 'experts' and 'simple respondents' are *de-emphasized* so that generalizations about public opinion can be induced from responses given by a small number of select interviewees. It is argued that such a survey can provide a 'sounding out' of public mood, and that while it 'does not fully or adequately reflect the state of mass consciousness, it allows [one] to speak about tendencies in social public opinion' (Isaev and Ibraeva 1995; see also Isaev, Ibraeva and Madaliev 1995). Similarly, Isaev, Shaidullieva and Madaliev (1998) have asserted that even competing responses from experts are valid in so far as their opinion 'expresses [that] of the masses' and hence reflects the diversity of opinions within broader publics.

In each case, the difference between 'experts' and 'simple respondents' is reduced to a quantitative question of whether generalizations about larger populations can be drawn from the responses of a smaller and more purposefully chosen segment; a choice made for financial reasons, but which is nevertheless made acceptable within the bounds of acceptable sociological methods. The legitimacy of 'expert surveys' and of the political interpretations made on the basis of them is therefore justified by a theory of relative representativeness.

In other cases, however, differences between the knowledge of 'experts' and 'simple respondents' are *emphasized* to explain why surveys produce disparate results (Isaev *et al.* 1993, 1994a; Isaev, Shaidullieva and Madaliev 1998). In these cases, sociologists argue that experts and non-experts are two fundamentally different species of knowers. Thus, because their responses cannot be expected to be similar, inconsistencies in the results of studies conducted among 'experts' and 'the masses' do not challenge the validity or reliability of the studies themselves. This was explained in an article published on political ratings in April 1994 (Isaev *et al.* 1994a), worth quoting at length because it offers clear insight into the logic behind these rhetorical distinctions.

> All residents of Kyrgyzstan who are included in a [sampling frame] may be respondents, depending on their sex, age, region and place of residence, nationality, social means and other indicators, depending on the goals of the study.... By correctly creating the sample, we can guarantee that the opinion of our respondents generally reflects the opinion of the whole population.... When we conduct a survey of experts, however, we are already not asking simple residents of the republic, but specialists and professionals who study an area of social life that we are interested in. In this research, political scientists, sociologists, psychologists, journalists and activists from political parties are all experts. For objectivity, we select expert personalities who do not work for power structures. For example, in the President's Apparatus, the *Jogorku Kenesh*, the government and oblast *akimiats* [councils] there are

plenty of professionals who have candidate and doctoral degrees, but for fully understandable reasons we do not invite them to be experts in our studies. As far as an expert survey is a survey of professionals, its results have a greater degree of prognosis, because professionals, in contrast to simple respondents, are obliged to have a broader perspective. Therefore, we think the results... do not contradict one another.

There are several interesting themes in this passage, including concern about the politicization of scientific knowledge and its effect on sociological method, and the distinction made between the legitimate professionalism of 'activists of political parties' and the illegitimate work of 'personalities who... work for power structures', including academics holding positions in offices of state and regional administration.

Most interesting, however, is the boundary-work done to elaborate two different but equally valid types of 'objectivity' in sociological research, one for studies that subjectify the 'masses', and one for those that subjectify 'elites'. Each draws its legitimacy from a different epistemological source. The validity of mass surveys is contingent on statistical probability and representativeness, while that of an expert survey is determined by the 'individual character, intellectual and professional level, and propriety and honesty' (Isaev et al. 1997, 1997b) of respondents and the belief that professionals will have more accurate and objective views on social and political affairs (Isaev et al. 1994a; see also Isaev et al. 1996a). It is also significant that expert surveys are more frequently associated with good sociological practice western countries (Isaev, Shaidullieva and Madaliev 1998).

This is a question of competing philosophies of truth (e.g. representation versus interpretation), but based on normative hierarchies of knowledge and ontological assumptions about knowers. As a result, different criteria of objectivity may be ascribed to different survey methods. There is little discussion of how the problem of representativeness may be addressed or challenged through the use of expert surveys, how political affiliation or qualities of 'propriety and honesty' may also affect the answers of 'simple respondents', or how researcher bias might influence the definition and identification of 'objective' and 'honest' experts in the first place. Nevertheless, these debates are excellent examples of how the 'selection of one or another description depends on which characteristics best achieve the demarcation [of sociology] in a way that justifies scientists' claims to authority or resources', and good illustrations of the argument that science 'is no single thing: its boundaries are drawn and redrawn in flexible, historically changing and sometimes ambiguous ways' (Gieryn 1983: 781).

Mapping the field: delineating science and politics

In addition to concerns about representativeness and validity, the research group was heavily criticized for the way it constructed lists of political leaders.

Controversies over research design reached an apex when in 1994 the BHU group removed president Akaev's name from the list of politicians being rated (Bakir Uluu 1997; Bekturganov 1994a). Though Isaev was accused of 'shuffling the data', he claimed it was an attempt to diffuse political debates, which had emerged after previous ratings surveys suggested that the president's popularity had declined (1996a). More specifically, he argued it was a way to 'consolidate society, avoid conflict with individuals and develop sociology in [the] republic' (Isaev *et al.* 1994b). Pro-government critics, however, interpreted the decision as a politically motivated attempt to symbolically exclude Akaev from the political landscape and criticized it for being a 'distortion of the principles of correctness for the selection of experts, methods, techniques and procedures defining political ratings' (Bekturganov 1994a; also Bakir Uluu 1997).

Bakir Uluu's (1994, 1997) accusations of researcher's bias were rooted in deeper concerns about the effects of political, and in this case specifically party, bias. They were a response to Isaev's (1993b) claims that Bakir Uluu himself had conflated social scientific criticism with 'party work'. Both men are prominent academics who also hold positions of responsibility in political (opposition) parties in Kyrgyzstan; both, therefore, are forced to justify dual affiliations to science and politics. The heteronomous relationship of social science and the Communist Party during the Soviet period has made this a particularly contentious issue in post-Soviet Kyrgyzstan. During perestroika, questions of scientific authority could still be pursued within a context of political commitment and often from within the Communist Party (Admaliev and Tuzov 1991; Blum 1990; Isaev 1991a, 1993a). After independence, however, the mere association of sociology with any political party makes academic work liable for labelling as illegitimate, politicized, unscientific and immoral. Bakir Uluu (1994) reinforced this clear boundary between scientific and political activity when he published an open letter to Isaev, saying, 'let's be ethical about this question: both you and I, aside from our party work, work in institutions of higher education. We cannot separate one from the other. It is a different matter if you ... want to reintroduce principles of the party into science'.[14] Isaev, when confronted with questions from sceptical journalists about the possible conflation of truth and power as a result of this dual affiliation, replied,

> the ultimate treasure for an academic person, a scholar like me, is the independence and freedom of his country where scientific justice is upheld as a priority. Few who value this treasure. Opposition and opponent are not to be confused, for they carry absolutely different meanings. I consider myself a patriot–opponent.
>
> (Isaev 1998)

Hence, he justifies both intellectual legitimacy and social relevance by distinguishing his academic and political work and defining the former as a moral imperative of the latter.

In such debates over the science–politics boundary, there is little concern that social science will have an adverse effect on political work; in fact, it is believed to increase the transparency, effectiveness and justice of socio-political action. However, there is also no sense that ideological and political affiliations, beliefs or practices can play productive roles in sociological work. Central Asian sociologists thus recognize two legitimate articulations of social science and politics: either an unambiguous separation of the two or a didactic relationship in which sociology informs political action but political action has no effect on sociological work. This is commensurate with two dominant theories of knowledge: first, that sociology can and should contribute to the development of scientific politics, and second, that one can maintain clear boundaries between 'science' and scientific knowledge and 'politics' and political knowledge in the process.

The location of the boundary between these two fields is drawn along a normative axis of intent, and grounded in positivist theories of knowledge that ascribe a negative role to subjectivity in social science and correspondence theories of truth which eliminate the agency of the subjective knower from the production of statements about social reality. 'Good' knowledge production may have political implications and remain legitimate as long as it objectively reveals the status quo of 'social reality'. However, when it is perceived that 'the research programme itself serves to corroborate a priori ideals born of someone's political ambitions', it crosses into the realm of ideology (Blum 1991). Despite this fear of subjectivity on the validity of truth, some allow that 'any concrete sociological study, whether it is objective or not, has a certain level of subjectivism' (Isaev et al. 1994a). The authority of social scientific knowledge, therefore, depends on the extent to which the causes of this subjectivism can be practically or rhetorically eliminated.

Two common strategies for achieving this among Kyrgyzstani sociologists are to create spaces of intellectual autonomy within the political field and to use positivist empirical methods in social research. In both cases, practices intended to guarantee neutrality and objectivity are opposed to those founded on politicization and intent. This is illustrated in a sample of text that appears in several articles on political ratings (Isaev and Ibraeva 1996; Isaev et al. 1994a,b; Isaev, Ibraeva and Madaliev 1995):

> we strive to conduct our research independently from the power structure and various political forces in the country. For members of our independent sociological group, objectivity, scientific conscientiousness and the quest for truth, as well as the observation of widely accepted methods of conducting applied research, are obligatory concepts.

The meaning of 'objectivity' here is ambiguous and contextual. There is no central group or institution that represents a legitimate consensus on its definition; no agreed upon set of disciplinary guidelines. Questions about 'who will judge' the difference between 'real' and 'pseudo' social science (Isaev 1993, 1994b) are fertile ground not only for boundary-work about methodology and the relationship between science and politics but also for negotiations about the professional ethos of sociology in Kyrgyzstan.

'Clean hands and clean minds': constructing a new ethos of science[15]

Methodological standards are moral as well as technical imperatives of social scientific practice. According to Merton (1996[1942]: 267), an ethos of science is 'that affectively toned complex of values and norms which is held to be binding on the man [sic] of science'. They not only shape the meaning of science within different historical and cultural contexts, but also influence the 'scientific conscience' of practitioners. As such, 'the mores of science possess a methodological rationale but they are also binding, not only because they are procedurally efficient, but because they are believed right and good' (Merton 1996[1942]: 268).

The professional ethos of sociology in post-Soviet Kyrgyzstan is what enables academics to distinguish between 'good' and 'bad' knowledge and practice, and which, like other elements of the field, is contested. Its establishment has been a highly contentious process as different groups of practitioners struggle to institutionalize different definitions of 'good' sociological practice, and as new alliances between sociology, the state and international and commercial organizations come into conflict with existing norms regulating the relationship between academic and political work. New ethico-moral values such as disinterestedness and non-commerciality that have been embraced in theory are difficult to sustain practically in conditions where knowledge is judged according to its degree of political relevance and is almost entirely dependent on external sources of funding.

This tension is also historical. While Soviet sociologists adhered to or were bound by Marxist–Leninist norms of professional practice, strains between the logic of science (their subjective loyalty to the pursuit of scientific truth) and their voluntary or requisite subordination to the logic of power created an ambiguous ethics. By the 1990s, 'good' social science could be defined in two ways, both of which were distinguished from mainstream western theories of 'modern science' in their rejection of disinterested objectivity and their encouragement of political commitment in scientific work (see Merton 1996: 274–76 on 'disinterestedness' in modern science, and Inkeles 1958: 138 on Marxist criticisms of 'objectivity' in intellectual activity). The first, produced by academics that supported party dominance in the academy, emphasized the political use-value of sociological work. Officially, good sociology in Soviet Kirgizia was that which met the administrative and ideological needs of the ruling regime. The second definition of 'good' sociology, elaborated less systematically and more discretely by those critical of political control over intellectual activity, was also based on notions of political-use value in that 'good' research was that which could be employed in the pursuit of socialist ideals. The difference, while subtle, is that the latter definition privileged scientific over political truth-claims and personal morality over formalized ethics. In this discourse, 'good' knowledge created foundations for political action and not vice versa; 'good' political power was that which was subordinated to scientific authority.

These two conceptions continue to influence the new professional ethos of sociologists in the post-Soviet period, which is developing in response to changes in the epistemological foundations and cultural organization of social science. This ethos has five major elements, each of which is invoked throughout the political ratings articles. First, social scientific work must be motivated by purely 'scientific' intentions which are not 'corrupted' in any way by personal or political aspirations to power. This norm is expressed, for example, in Bakir Uluu's arguments that his research is more legitimate than Isaev's because his goal was 'the search for truth and not power', a claim imbued with normative force by his symbolic association of his own research with the 'European school of education' and his opponents' with national patriarchal traditions (1994, see also 1997). This is a clear departure from Soviet conceptions of *partinnost'* [loyalty to party line] and a shift towards the notion of 'disinterestedness' which grounds positivist ideologies of modern science in the west.[16]

This is connected to the second element of the emerging ethos, which is that research should be non-commercial and not for profit, conducted to inform decision-making, enlighten the public and advance knowledge in general. Both Bakir Uluu and Isaev invoke this norm to challenge the authority of one another's research. While Isaev protested against the employment of sociological research as 'political prostitution' (Isaev *et al.* 1997a), Bakir Uluu accused him of betraying the field through a 'marketization' of science (Bakir Uluu 1994). This again reflects a move away from conceptions of applied social science as a technical service to the power elite and towards ideals of autonomy and informational 'communism' which characterize dominant ideologies of science in capitalist societies (for more on communism in science see Merton 1996: 271–74).

Third, sociological work must be conducted in the most 'objective' way possible, with every possible influence of subjective interpretation being accounted for at every stage of research. This runs contrary to previous norms, which defined such objectivity as a 'bourgeois' ideology that prevented the recognition of social inequalities and injustice. Fourth, sociologists must adhere to the scientific method; it is argued that the superiority of this approach to that of 'speculative philosophy' is demonstrated by advances in Western social science. Finally, sociologists must occupy a moral high ground by being transparent about the limitations of their knowledge and motivations for their research, and by not allowing themselves to be influenced by extra-scientific forces. Engaging in political debate, even about sociology itself, is frowned upon and portrayed as 'uncharacteristic' of social scientists; however, it is justified if such debates are deemed necessary for defending the moral superiority of social scientific work over political truth-claims (Isaev *et al.* 1994b). Sociologists who do engage in debates with one another over the legitimacy of particular studies, methods or practices therefore often preface their political arguments by reaffirming their commitment to the principles of intellectual autonomy and professional morality. 'Integrity' is highly valued in this discourse, where absolute objectivity has become a primary criterion of scientific legitimacy despite acknowledgements that it may not actually be a practical possibility (Isaev, Ibraeva and Madaliev 1995). As Isaev argued in one response to Bakir Uluu, 'it is doubly important if we consider that sociology in our republic is going through a growing phase. For members of our sociological group...researching socio-political and other processes in Kyrgyzstan, honesty and objectivity, as well as scientific laws, are sacred values' (Isaev *et al.* 1993a, 1994a).

Notions of honesty, integrity, morality and rational action also play a role in establishing scientific authority within the public sphere. They are familiar as they are also prevalent in more general discussions about post-Soviet social ethics in Kyrgyzstan and resonate with a public that feels it lives in a 'period of global transformation of consciousness and a deep break of norms and behavioural stereotypes' (Bekturganov 1995). Sociologists who do not conform to these new norms or who are judged by others to be in violation of them are often labelled 'pseudo-sociologists' by their peers and excluded from the academic community (Baibosunov 1993; Isaev *et al.* 1993a, 1997a). However, the inherent ambiguity of the broadly agreed upon terms of reference means that such labels may be assigned arbitrarily and, in many cases, for deeply political reasons.

More importantly, the dominant professional ethos of disinterestedness (purity of intent), non-profitability (communism in science), objectivity, scientificity and moral integrity, which is elaborated in the articles on political ratings, is often subverted by the existential realities of academic life in Kyrgyzstan and by the hierarchical, competitive organization of sociology in the republic. It is also complicated by the dual-pronged project to establish scientific legitimacy and social relevance. Appeals for methodological and social disinterestedness in the pursuit of sociological truth exist in tension with demands for sociologists to produce politically relevant research. The commercialization of research has made researchers vulnerable to criticism, as they are heavily dependent on contract work and commissions. In the absence of centralized academic standards and in institutions where corruption is rampant even in the highest echelons of the academy, professional competition often makes the maintenance of professional norms itself part of a symbolic struggle. The ideal of free information within the scientific community is unattainable within this environment, in which knowledge and expertise have become commodities and potential sources of social capital and professional power. While adherence to the procedures of scientific method is marginally easier to evaluate, disagreements about what constitutes an 'acceptable approach' make this a contentious area as well.

In other words, although Kyrgyzstani sociologists are constructing a professional ethos of science that aims to transcend structural constraints and mediate contradictions in their relationships with one another and with other social institutions, it has not yet become institutionalized as a professional code of practice. Instead, it remains most effective as a rhetorical device with which professional norms can be further elaborated, debated and contested. The tensions in the ethos of Kyrgyzstani sociology are central to understanding why, nearly five decades after its initial emergence in the republic, sociology in any form has not been institutionalized as an academic discipline or professional practice. Many setbacks – the censorship and reorganization of the KSU laboratory, the repression of research on ethnic relations, the lack of support for sociological work, the inability to establish indigenous sociological institutions, the lack of material and symbolic resources, and the breakdown of productive relationships within the academic community – are results of structural constraints such as authoritarian government, centre–periphery inequalities, intellectual and financial poverty and academic dependency. Theories about the state of the discipline that focus on these factors, such as those introduced in Chapter 1, are therefore not misguided. However, they are incomplete. These and other problems are also created by non-material factors, particularly the ways in which the nature and role of sociology, as well as the meaning of its reform or 'transition', have been conceptualized and articulated by Kyrgyzstani sociologists themselves.

CONCLUSION

The project to transform social science from a way of knowing that defines 'truth in strength' to one that can assert 'strength in truth' epitomizes the intellectual zeitgeist in late socialist and post-Soviet Central Asia. Public discourses on science are rife with talk about autonomy of mind and nation, purity of deed and intention, transparency and truth; these are regarded as moral imperatives in societies where the discrepancy between ideological and lived reality continues to complicate the cultural meaning of 'truth'. Ideologies of science not only play an important role in resolving this tension, but also obtain in these conditions the epistemological authority necessary for their development and legitimization. Whereas in the Soviet Union such rhetoric was rooted in the ethics and politics of dissent, during the post-Soviet period it has become embedded in the discourses and institutions of global democracy and development. And yet, within global capitalism as Soviet communism, social practices in projects to institutionalize the field remain remarkably similar, the most fundamental problematic being the relationship between knowledge and power, variously defined.

The many different projects to reform and institutionalize sociology in Kyrgyzstan from the mid-1960s to the present day tell one version of this story. They have in common an underlying desire to divorce the production of truth about social life from the exercise of political power; to wrest the power of truth from the hands of those perceived to employ it to maintain illegitimate types of power; to change a heteronomous field of knowledge and practice into an autonomous one. The presumed solution to this problem is the creation of conditions in which people – political leaders, citizens and academics alike – can seek social truths outside the logic of power. Acquiring 'strength in truth' can only be achieved, it is argued, as long as the quest for social reality is pursued in isolation from personal and political interests, particularly through empirical observation conducted according to the scientific method. Truth-claims constructed in this way are believed to be objective, politically neutral and value-free and useable guides for social and political action, thus forming the foundation for a rational scientific politics that can stem domination by asserting the strength of legitimate truth.

This movement has been interpreted in two ways. One is as a movement towards democracy and 'world' science. In this context, the development and

145

institutionalization of a scientific sociology that functions to preserve truth and justice amidst ideological distortion and domination is in line with the traditional philosophy of social science – but also, it can be argued, the moral imperative underlying much critical social scientific work. The second is that the attempt to institutionalize scientific sociology in Kyrgyzstan represents a new type of naïve positivism, brought about by years of intellectual repression, Marxist brands of epistemological reductivism, and cultural tendencies towards determinism. Both interpretations are underlain by anti-communist rhetoric, narrow and ethnocentric understandings of the history of sociology, Orientalist assumptions about the superiority of 'Western' knowledge and knowers, and a general lack of information about or regard for non-Western academics. They are also reinforced by the exclusion of important insights from the sociology of knowledge and science which have in recent decades helped to illuminate the more general politics of organized knowledge, particularly regarding the quest to make 'scientific' forms of knowledge which are inherently situated and political.

By asserting that 'belief in the value of scientific truth is not derived from nature but is a product of definite cultures' (Weber quoted in Merton 1996 [1938]: 277), however, it can be argued that the development of different conceptualizations of sociology in Kyrgyzstan is neither an inevitable consequence of the Soviet collapse nor an indication of intellectual inadequacy, but a result of cultural work within the academy. It may be understood as a deliberate, if often under-theorized, attempt to reorganize the relationship between knowledge and power in order to democratize both and pursue professional agendas.

At one level this is a simple equation: good science must be socially useful. However, definitions of 'good science' and meanings of 'relevance' are dynamic and political. During the 1970s and 1980s, for example, good sociological practice was Marxist; linked to the realization of Soviet socialist ideals of justice and equality, as well as to modernization and industrialization as articulated in party ideology. Later during perestroika, Marxist sociology continued to pursue these general goals but was reconstructed as a critical counterweight to the authoritarianism and ideology which sociologists argued had 'distorted' original socialist agendas. It also gained a new role as part of efforts to increase the autonomy of peripheral republics within the Soviet empire, thus becoming integrated into movement to articulate non-Russian national identities. After independence, adherence to Marxist theories and ethics no longer constituted good practice, but in fact became the criterion for a new type of 'pseudo-sociology'. Embracing what Marxism had rejected – positivism, empiricism, epistemological 'objectivity' – became central in projects to reconstruct the field's identity. The institutionalization of non-Marxist, market-oriented and publicly relevant sociology became a symbolic indicator of the rationalization of political power within the society.

For the academics involved, these obviously constitute very different, perhaps even mutually exclusive sorts of projects. However, they may also be seen theoretically as different manifestations of a common exercise: in each case, after separating the production of sociological knowledge from the logic of illegitimate power in order to establish scientific legitimacy, sociologists have needed to associate its application with that of legitimate strength to promote the discipline's social relevance – and, in the process, delineate boundaries of legitimacy and authority around both knowledge and power. In particular, they engage in three types of boundary-work: erecting rigid borders between social scientific knowledge and power at the level of knowledge production; blurring boundaries between scientific knowledge and power at the level of knowledge application; and articulating ideals of either scientific politics (in the professional–applied model) or enlightened activism (in the liberal–critical model) which naturalize the combination of logics in the two activities. These categories and relationships, created in order to order, institutionalize and promote the discipline, have been naturalized by academics themselves. They have become integrated into new sociological concepts.

While the discourse of knowledge production in Central Asia is anti-political, academic and intellectual practices are highly politicized. The research problems that dominate Central Asian sociology today are no less over-determined by forces external to the discipline than those which prevailed during the Soviet period. They are formulated in response to public opinion about social problems, dictated by the administrative and ideological needs of states and capitalist markets, and consumed by foreign *zakazchiki*. While the drive for autonomization thrives in the scientization of sociological method and ethics, intellectual content remains heteronomous. Although the nature and role of social science is no longer dominated by an imperial state, the abrupt end of intellectual colonialism, including its financial subsidies, engendered a new type of academic dependency.

There is thus an inherent tension between new scientific discourses of sociology and new practices of knowledge production currently being institutionalized in Kyrgyzstan. While the most immediate problems for practitioners may be the repoliticization of organized knowledge or the resurgence of positivist or technocratic social science, the main problematic from a theoretical perspective is that the goal of creating an autonomous academic discipline that may develop in greater freedom from the dictates and logic of political power is contradicted by two things. One is the predominance of epistemological assumptions and methodological practices which are themselves politicized and which proscribe the analysis of the actual relationship between knowledge and power within the field; the other is the introduction of new forms of economic and political heteronomy which are rhetorically represented as avenues towards freedom.

In other words, while the politics of knowledge in Kyrgyzstani sociology are at one level localized manifestations of one professional community's attempt to come to terms with the collapse of empire and reconstruction of society, they are also part of larger questions about the fate of academic work, the role of social science, and the political economy of 'truth' under conditions of postsocialism. They force us to reflect critically upon how our own hegemonic conceptions of truth and 'good' social research have been articulated and contested; what social conditions have enabled these formulations to emerge; and under what conditions we too might seek out alternative epistemologies. It returns us, in other words, to the issue of power/knowledge in science itself. In addition to asking what the state of social science says about Central Asian modernity or development, or inquiring into how Central Asian academics have delineated boundaries between sociology and politics in their everyday lives, we may also ask what this says about our own understandings of the relationship between realism and constructivism, and between truth and power. Is the quest for autonomous truth about society still a valid goal for social scientists to pursue? Can it be achieved without disregarding the contributions of critical theory regarding the dialectical relationship between subjectivity and objectivity in social scientific research, and post-structuralist insights into the all-pervasive reach of power in knowledge? Can theoretical work about the politics of knowledge contribute practically to the development of a politically engaged and yet scientifically autonomous discipline? If so, could sociological practice in societies such as Kyrgyzstan disrupt the prevailing power/knowledge *doxa* that often divides positivist and critical sociologists? Is there still place in social science – in the former Soviet Union and elsewhere – for an epistemological belief in the cultural and political value of truth? The experiences of Central Asian social scientists demand that these questions be taken seriously. They demonstrate on one hand that intellectual colonization, domination and dependence are intolerable even for academics invested in these very relations of power, and on the other that attempts to deny the intricate relationship between knowledge and power in social scientific work are untenable even for those invested in the ideology of scientific objectivity. Conditions for the creation of autonomous knowledge, for the production of 'truth' about society, have not been created within Soviet socialism, post-Soviet capitalism, nationalist authoritarianism or 'international civil society'.

One conclusion that may be drawn from this is that belief in autonomous truth is itself a modernist myth that should be abandoned. While such critical postmodern philosophies of knowledge have definite value in the Central Asian case as necessary critique, they must also be able to contend with difficult questions about the role of knowledge in advancing human freedom. This is yet another Enlightenment agenda, but one that has daily ontological implications for those living in subjugated conditions and which cannot be easily dismissed even within constructivist philosophies of knowledge. An alternative conclusion is that knowledge production may yet be made more valid and autonomous under the 'appropriate' conditions, and that these do not exist in Central Asia. Critical studies of how social science might be scientific, however this is defined, and empirical without being positivist and empiricist might contribute to such an approach – though understandings of what would guarantee the 'social conditions of the possibility for rational thought' will need to be considerably re-theorized in the postsocialist context, and insights from the sociology of knowledge and science must not be ignored. Finally, we might simply conclude that more explicit, theoretical and reflexive work in the sociology and philosophy of knowledge and the organization of social science within postsocialist space is necessary to engage these questions at their roots, and to elaborate new ways forward beyond the confines of Marxist, nationalist and capitalist orthodoxies in social science.

POSTSCRIPT

Throughout the process of this research, many colleagues asked me how it would ultimately benefit them and were generally unsatisfied with my responses. This tension between a struggling professional community seeking informed solutions to practical problems and my more theoretical curiosities and ambitions is part of the broader debate surrounding the role of sociology. It has also been a discomforting reminder that I have not fully transcended the divisions and power relations discussed at the beginning of this book. I, after all, have had the privilege of publishing it, which I fear in some senses makes me a willing accomplice in maintaining an entrenched colonial apparatus. However, the other logical conclusion of this critique, not to publish at all, fails to resolve the dilemma. The experiences of Central Asian academics matter for the broader sociology of knowledge and history of science, which, despite years of corrective measures, remain largely Eurocentric in content and methodology. They matter for our understanding of the Soviet and postsocialist experiences, particularly in so far as Central Asian societies continue to be marginalized both politically and academically. They matter for the study of global capitalism, whose contradictory cultural effects are felt and mediated so diversely throughout the third world. They help to clarify, without answering categorically, questions about how we might understand Central Asian societies in the context of colonialism, nationalism and globalization and the political economy of truth in a postsocialist age. Censoring the work, therefore, seems equally as wrong, if not more so.

And yet, this is another inadequate response to those who are waiting to reap benefits from contributing to this research. It cannot offer what many people immediately want and need: more money, more freedom and improved access to resources. Action research building on it might be able to make advances in this regard; ultimately, such problems will be resolved politically, through human agency, in ways similar to those discussed throughout this book. Research, after all, can constitute only part of such projects. However, this research assumes that historical and theoretical insight can inform individual and collective praxis, and that this would be its primary contribution. I therefore have four supplementary suggestions for how this work might be engaged. The first is that by removing Central Asian academics from the confines of area studies and situating them within the broader history and sociology of knowledge, it becomes more possible to draw on comparative research which has both theoretical and practical implications for analysing and informing contemporary practice. The second, related proposition is that it demonstrates how the practice of social science within the region, even at the level of departments, may also be understood and engaged at the level of wider social and economic forces. Third is that this analysis of the politics of knowledge in Central Asia offers a coherent set of theoretical positions to which Central Asian academics may respond with their own alternatives. In order to reduce or eradicate inequalities within the global science system, we need what Smart calls an 'interpretive' rather than a 'legislative' international sociology, 'one which attempts to offer a translation service between different cultures and communities' (1994: 158). Studies such as this, as well as comparative studies or rebuttals that may follow from it, may be one way of advancing this project. Finally, it illustrates how the foundational questions of sociology, particularly its political and scientific statuses and the effect of these on its intellectual legitimacy and social relevance remain crucially important in the formation of the field. As questions and potentialities, they also lie at the heart of truth-claims in all social science. It is thus a reminder that at any given moment, the boundaries between science and politics, truth and power in social scientific context should be seen as just that, and that consciousness of this ambiguity should be a stable presence in our ordinary and extraordinary knowledge practices.

NOTES

PREFACE

1 I was based at the American University in Kyrgyzstan (now AUCA) and had additional responsibilities in Kazakhstan and Uzbekistan. CEP was 'founded in 1991 [as] a private, non-profit educational organization . . . to educate a new generation in Central and Eastern Europe and Eurasia in the principles and habits of democracy. Believing that critically minded and informed individuals are fundamental to a thriving democratic society, CEP work[ed] with universities throughout the region to bring Western-trained social science academics and lawyers to their institutions' (CEP 2000: 203). The programme was withdrawn from Central Europe and Eurasia in 2003, though a number of others continue to operate through the Open Society Institute.
2 See Bruno (1998) for a critique of the practice of 'training' by foreign NGOs in the former Soviet Union.
3 Kalb (2002: 320) has written a useful analysis of how 'globalization' and movements to establish 'civil society' in the region have disempowered those from marginalized groups and regions, as well as increased stratification within communities through things such as 'brain-drain' and the 'foreign-imposed strengthening of one cluster of political actors over another'.
4 The term 'epistemic negotiation' is borrowed from Reeves (2003a) and reflects a more symmetrical understanding of the way in which knowledge structures and power relations underlie what some anthropologists refer to as cultural 'misunderstanding'.
5 For more on the importance of 'situating ethnography within its historical and geographical context', see Burawoy (2000: 25).

INTRODUCTION

1 For other calls to situate Central Asian Studies in broader theoretical frameworks see Khalid (1998) on Central Asia as part of broader Islamic history and Cavenaugh (2001), Moore (2001) and Verdery (2002) on the postcolonial perspective.
2 For an interesting critique of 'Marxism–Leninism' as an 'empty signifier' which is actually used as a rhetorical device for ideological control but which also allows for some interpretive liberty (see Walker 1989).
3 The significance of this has become increasingly visible in countries such as Uzbekistan and Turkmenistan, where the censorship and exclusion of foreign organizations has compounded the deterioration of education and intellectual life (see Bartlett 2006; Megoran 2006; Mikosz 2001).
4 Articles have recently begun to appear on regional differences in Soviet sociology (see Zborovskii 2001).

1 KNOWLEDGE AND POWER IN
POST-SOVIET SPACE

1 In an attempt to identify the centre or 'core' of international sociology, Alatas (2003: 602) defines 'the West' as 'the contemporary social science powers, which are the United States, Great Britain and France'. The definition in Kyrgyzstani sociology, however, is more fluid (Blum 1993; Fanisov 1990). In some cases 'the West' includes Western Europe (and/or the United States, and may refer to Russia (Isaev 1998b; Isaev *et al.* 1997a; Ismailova 1995). However, 'East' and 'West' are also employed symbolically. There are two main symbolic uses of 'the West': the West representing civilization, rational progress, modernization and order (e.g. Isaev 2000), and the West that represents moral anarchy, pornography, consumerism and excessive individualism (e.g. Isaev, Akmatova and Dosalieva 1996). Likewise, there are two meanings of 'the East': one representing personalized power, tribalism, patriarchy and backwardness, and one symbolizing national purity, pre-colonial identity, indigenous knowledge and collective humanism. For a Kyrgyzstani perspective on the difference between 'Eastern' and 'Western' conceptions of 'open society' based on Gandhi and Popper respectively, see Isaev (1998d).

2 It is important to recognize that epistemological debates within European and Russian Marxist philosophy and philosophy of science were vibrant prior to the Stalinization of intellectual life; see Gasper (1998) and Sheehan (1993). Lenin (1908) vigorously defended a dialectical and materialist epistemology over the phenomenological realism of scientific knowledge in early twentieth-century debates with Russian Machists; the ascendance, institutionalization and modification of this philosophy after the Bolshevik revolution, rather than his later, more nuanced and more constructivist writings, was significant in the development of understandings of 'Marxist' and 'Marxist–Leninist' epistemologies later throughout the Soviet Union.

3 Mannheim's quote has been maintained in the original English translation despite its use of gendered language.

4 Kalb (2002: 321) describes the 'globalist grand narrative' in the former Soviet Union as one in which free-market trade engenders individualism and the development of a politically engaged civil society, and argues that this has 'substituted for the complexities of ethnography a simple self-reinforcing spiral of historical causation and moral teleology from markets to individualism to civil society to democracy to prosperity' in ways which are counterproductive to development.

2 THE COLONIAL LOGIC OF SCIENCE
IN CENTRAL ASIA

1 For similar definitions of academic or scientific colonialism see Bujra (1994), Lamy (1976) and Rahman (1995); for a critical consideration of the relationship between development aid and knowledge production see Doty (1996).

2 Khalid makes a persuasive argument for maintaining a dynamic and non-dichotomous theory of the colonial encounter in his book on Jadidism in Central Asia, noting that simple categories of 'Russification' and 'Westernization' often 'render themselves incapable of exploring the politics of cultural production and identity formation on the Russian borderlands' (1998: 15). This, I argue, is as true in the post-Soviet period as it was in the pre-Soviet – though it does not automatically imply that Russification or Westernization do not occur.

3 'Jadid' means 'new' in Arabic and in this case referred to a radical new text-based method for teaching literacy in Muslim Turkistan which was developed during the late nineteenth century. For more, see Khalid (1998).

4 For example, during the nineteenth century, Central Asian historians produced critical scholarship about Turkistan, particularly the history of Islam. However, those educated in newly established Soviet universities in Russia in the early twentieth century re-branded these narratives as reactionary, and the Communist Party eventually banned the use of the name 'Turkistan' after the Turkistan ASSR was dissolved in 1923 (Allworth 1998: 70; Shahrani 1994: 64). At its most extreme, the censorship of Islamic history involved the physical destruction of Central Asian Muslim scholars, educational and social institutions, libraries and texts (Shahrani 1994: 65). Gradually, favourable or politically neutral references to Islam and all mention of 'national' events and heroes were abolished from Central Asian history, as were local styles of narrative, and replaced with Marxist–Leninist theories of historical development and Soviet events and figures. Soviet secularization began to incorporate 'scientific' theories about the relationship between Islam, feudalism and imperialism (Ro'i 1995: 18).

5 Research 'bases' were often the first institutions established in the non-Russian regions; in 1949, all bases were redesignated as 'filials', signifying a wider range of activities and greater integration into the national science system (Gaponenko 1995).

6 The relationship between *grazhdanstvo* (Soviet citizenship) and *natsional'nost'* (e.g. Kyrgyz, Uzbek or Tajik nationality) was a highly contested matter in 'nationalities' policy. See Shanin (1986).

7 The Soviet system of academic degrees progressed from *bakalavr* (equivalent to a bachelor's degree) to *magistratura* (master's) to *kandidatura* (similar to the PhD) and finally to *doktor* (conferred after completion of a second major dissertation). For convenience, I will use English variants of these terms throughout this book. Those that studied for a candidate degree will be called candidates, and those for a doctoral degree, doctorants.

3 THE SOCIAL SCIENCE PROJECT IN SOVIET KIRGIZIA

1 For more on competition from more 'marketable' disciplines see Reeves (2003: 11) and Raiymbekova, K. (1999: 51–56), cited in Reeves.

2 During this period, both the SSA and a section for sociological problems in the Moscow-based USSR Academy of Science (later the Institute for Concrete Social Research [1968] and then the Institute for Sociological Research [1972]) were established. In 1965, fledgling sociological groups at Leningrad University joined forces to form the Institute for Complex Social Investigations. The first meeting of Soviet sociologists, many of whom claimed to have been conducting sociological research for years, was held in Leningrad in February 1966 (Simirenko 1969: 393). The Central Committee of the Communist Party organized its own Academy of Social Sciences during this period, and government ministries, newspapers and industries began to commission studies on issues such as 'workers' discipline', time budgets and labour management. According to historians of Soviet sociology, 'by the mid-seventies no less than six hundred centres of one kind or another were said to be engaged in empirical work in 120 towns throughout the country' (Matthews and Jones 1978: 8).

3 In fact, a series of Soviet Sociological Association meetings in 1969 were held under the banner 'from dilettantism to high professionalism' in order to address this problem (Simirenko 1969: 394).

4 The concept of *natsional'nost'* ('nationality') has no precise corresponding translation in English. I have used 'ethnicity' because in this context the term referred to what might be considered ethnic groups ('Kyrgyz', 'Uzbek', etc.).

5 Others attribute this to Rakhat Achylova, who defended a thesis in sociology in Leningrad during the Soviet period (Asanova 2003).

6 The production of hydroelectricity, specifically for military use, was one of the Kirgiz Republic's specialized functions in the Soviet economy, along with shepherding and wool and cotton production. See Dabrowski *et al.* (1995).

7 'Propaganda' referred to ideological education within the Communist Party, while 'agitation' was more related to mass ideological work, particularly in mediating party policy and mobilizing workers (Inkeles 1958: 41).

8 The Communist Party established a 'council for the coordination of scientific research' in 1968, organized a council for the 'coordination of research in the sphere of historical-party sciences' in 1975, decided in 1976 to 'strengthen mutual ties of social, natural and technical science' with inter-sector, 'complex' research, and issued a decree on 'increasing the effectiveness of scientific research in higher educational institutions' in 1978 (Tabyshalieva 1984).

9 Translation compliments of Nienke van der Heide.

4 KNOWLEDGE AND NATIONAL IDENTITY DURING PERESTROIKA

1 Public 'criticism and self-criticism' were part of the socialization of Soviet morality and the creation of group consensus and self-censorship (Bronfenbrenner 1969: 290). In Kyrgyzstan, during early perestroika, this took the form of enforcing conformity, 'naming and shaming', and discouraging nationalism in social science. For example, in his article on the 'highest mission of sociology', Sherstobitov (1987: 4) criticized Kirgiz social scientists for writing revisionist histories that did not acknowledge the class-based nature of social conflicts and for being seduced by the practice of writing in 'idyllic tones' about 'reactionary-nationalistic and religious survivals', all of which were 'against...ideology, the socialist way of life, and the scientific world view'.

2 See also Shalin (1990: 1020–25) for a discussion of how Soviet sociologists renegotiated the balance between scholarship and advocacy during this period, and Brym (1990: 213) on the 'ambiguous relationship to power' and 'ongoing tension between ideological commitment and scientific distance'.

3 Interested readers may consult Elebaeva, Dzhusupbekova and Omuraliev (1991) for further analysis of this event.

5 SOCIAL SCIENCE AFTER COMMUNISM

1 For Central Asian perspectives on how these conditions have manifested in and affected social science, see Abdyrashev (1994); Ablezova (2003); Asanbekov (2003); Asanova (2003); Blum (1991); Botoeva (2003); Fanisov (1990); Isaev (1991b, 1999d); and Nurova (2003).

2 See Bronson *et al.* (1999) for a full discussion of the structural, intellectual, personal and political challenges facing post-Soviet social scientists, and a report by the Council of Europe called *Social Sciences and the Challenge of Transition* (2000) for a slightly different variation.

3 State investment in education declined dramatically after Kyrgyzstan declared independence. Egorov (2002: 61) reported that government spending for research and development declined from 0.7 per cent of the GDP in 1990 to 0.14 in 1999; Glenady (1996) supports this with her figure of 0.18 per cent in 1994. According to the Kyrgyz National Statistics Committee, in 2000 the state allocated 3.1 per cent of its annual GDP to education – less than half the amount allocated in 1995 (Reeves 2003: 9). In 2001 the president of the Kyrgyzstan Academy of Science claimed that the 17 million soms ($345,000) allocated to the Academy by the government was 'not enough to

achieve good scientific results' (Radio Free Europe 2001). In 2002, a local newspaper reported that state universities actually received only 10–15 per cent of their expenses from the state budget (Osorov 2002). See also Sydykov (1995) on the need to establish a union of scientists to protect the interests of scholars.

4 To take a local example, salaries at AUCA began at the equivalent of $80 per month for full-time instructors, while higher-level faculty members such as chairs and co-chairs received up to $250. Sociologists teaching at BHU more simply said they 'earn very little'. One full professor reported that her combined earnings from teaching at three different universities totalled $150 per month; another said that she earned approximately $26. Reeves (2002: 26) reported that in 2002, a local newspaper put average the salary for a new university teacher in Kyrgyzstan at $14.60 per month. See also Aslanbekova (2001) and Reeves (2003: 10, 16).

5 Such arguments are marginalized in contemporary scholarship about former Soviet societies, but they are not uncommon in postcolonial critiques of intellectual imperialism. See Chekki (1987), Chomsky et al. (1997), Fisher (1990), Garreau (1988), Gendzier (1985) and Lamy (1976).

6 Genov (1989) distinguishes between 'weak' and 'strong' definitions of national sociology. The first refers to the 'specificity of intellectual and institutional development in a given national social and cultural context', while the latter means 'outstanding contribution to the development of world sociology'. In Kyrgyzstan, the term is used somewhat differently: it incorporates elements of both 'weak' and 'strong' definitions and adds to them a moral imperative of national service in the face of colonial power.

7 Isaev identifies the main 'directions' in sociology as being family sociology, cultural sociology, conflictology, national customs, social structures and institutions, the study of the elite, national reforms and the middle class (Baibosunov 1998). These fields, however, cover the range of topics addressed by well-known sociologists working within the state system of government universities, research centres and the Academy of Science, and exclude the interests of sociologists working in private universities and non-governmental organizations.

8 Most work in the field takes 'Soviet sociology' as its primary unit of analysis. When specific local institutions are mentioned, it is generally by way of narrating the institutional development of Soviet sociology. The importance of regional and local variation within Soviet sociology is intimated by critics such as Shlapentokh (1987) and Popovsky (1979) who looked at qualitative inequalities within Soviet science, and revealed more explicitly in post-Soviet analyses of the social sciences in individual former communist republics (Keen 1994; Toschenko 1998) and institutional histories (e.g. Boronoev 1999; Grigorev 1999). Beliaev and Butorin (1982) have theorized the role of institutional actors and power relationships in the development of Soviet sociology; however, there are still few resources on the development of sociology in the Soviet Union at the institutional level.

6 RE-DISCIPLINING KNOWLEDGE IN KYRGYZSTAN: ALTERNATIVE VISIONS OF SOCIOLOGY BETWEEN MARX AND THE MARKET

1 There is some ambiguity about the exact date that this department was established – 1989 (Osmonalieva 1995) or 1991 (Baibosunov 1998).

2 Foreign funding is only one source of extra-governmental income for state universities such as BHU. Bribery has also become endemic since the collapse of the Soviet Union. It is difficult to obtain reliable statistics about how many educators accept or demand payment for admission, grades and exams; however, for discussions of bribery in

Central Asia see Obychnyi prepodavatel' (2000) and Osorov (2002). Reeves (2003, 2004) offers a useful analysis of how this practice is legitimized by the prevailing sense that in a market economy educational performance is somehow segregated from intellectual labour, and by alternative understandings of what it means to be part of an educational 'market'; is also why some regard it as an ethical decision within a dysfunctional educational system that would revoke financial support from students who would otherwise do poorly.

3 For alternative conceptualizations of 'generation' among post-Soviet academics, see Bruno's (1998) and Egorov's (2002) more economic categories; Bourdieu's more power-based ones of conversion, succession and conversion may also be usefully applied (Swartz 1997: 124).

4 Some programmes for educational or research exchanges set age limits, often thirty or thirty-five, on applicants.

5 According to Tishin (1999: 6–7), these functions 'determine the significance and role of [sociology] in modern society life. They are precise answers to the question of why, for what, people need this science'. They are: 'theoretical–cognitive' ('to accumulate and synthesize knowledge; to strive to present the fullest picture of the structure and processes of contemporary society'), 'world view' ('to give a general representation of the world of people'), 'ideological' ('sociological research is often used in political struggles for either kindling or overcoming social tensions; sociological data is often seen as a means for stabilizing society; sociological concepts are, for various groups of people, tools for argumentation and struggle in preserving their interests and goals'), 'humanistic' ('expressed in the development of goals for social development, programmes for scientific-technological, socio-economic and cultural improvements in society; sociology can mediate the improvement of human life'), 'predictive' ('on the basis of data gathered, [it] can determine the prospects for the regularities of life and development of society'), 'communicative' and 'economic' ('studies the state and dynamics of economic life, that is, the living components of the social structure'), 'administrative' ('to elaborate and help realize social policies oriented either towards the hastening or inhibiting of the socio-economic development of the state, which cooperates in the hands of one political force to form a homogenous society, and in another to differentiating society into unequal socio-economic classes and groups'), 'critical' ('to warn politicians about deviations in the laws of the development of social phenomena and processes and the possible consequences of these violations'), 'applied' ('to participate directly in developing and even realizing various social rec-ommendations, projects and experiments'), 'informational' ('to give primary data about individuals and groups of people, their needs, interests, value orientations and motives of behaviour, about public opinion and concrete conditions and situations'), and 'activization' ('to form public opinion and induce groups of people to act as someone needs, as it is advantageous').

6 The Russian term *distsiplin* refers to a specific body of knowledge and skills related to a thematic specialization, for example, demography, social statistics, social anthro-pology or social psychology. The more general term *napravlenie* (literally 'direction' but also used in academia to mean 'field of interest') is more analogous to the English-language understanding of an academic 'discipline' such as sociology. The main function of the national standards for sociology is to encourage the reproduction of an emerging disciplinary canon; not one of hegemonic theorists and schools of thought, but rather of bodies of specialized knowledge and skills. This includes, for example, 'general sociology', 'methods and techniques of sociological research', the 'history of sociology', 'political sociology', 'demography', 'social statistics', social anthropol-ogy', 'social psychology', 'sociology of education', 'sociology of youth', 'sociology of deviant behaviour', 'social modelling and programming' and 'social structures of society' (BHU 1994, 1995a). Within this broad canon are dozens of sub-canons

detailing the specific authors, topics and skills that should be taught as part of particular courses.

7 The department has offered, for example, courses in the sociology of youth, comparative sociology, deviant behaviour, the 'sociology of the individual' and Kyrgyz culture. As younger instructors joined the faculty, they introduced additional courses on media, stratification, labour, marriage and the family, *konfliktologiia* and civil society (BHU 1995a, 1996, 1997a). The department has also been developing a new component in gender studies since 2002.

8 These figures are from 2003 departmental records, which specify the number of courses taught by each member of the department. Reeves (2004) estimates that university instructors in Kyrgyzstan routinely spend 500 hours in the classroom per academic year.

9 In the Soviet academic system, faculty working in state universities were expected to implement centrally produced *uchebnye plany* in all of their programmes. As Ryskulueva remarked, 'everything went through one Soviet Ministry of Education in Moscow. We were sent documents that we had to implement and deliver, and everything had to be done according to form: we either had to give them to people or transform and adapt them and then give them to people. ... They created them in Moscow [where] they had scientific institutes, large-scale administration, state structures ... and everyone in the Soviet Union simply had to approve ... these pre-prepared plans. Therefore, the [Kirgiz] Ministry of Education had no real influence on the development of sociology during this period. ... You could go anywhere and you would find that sociology was the same in all ... the republics of the Soviet Union'.

10 Many students, however, feel that the diploma is less significant than personal and family connections in informal power networks, which are often required to gain employment (personal communication with final year sociology students, BHU, 19 March 2003). See also Reeves (2004).

11 Because it is bestowed from without, however, this legitimacy is tentative and must be vigilantly maintained. In the 2004 version of the national standards, for example, the status of sociology shifted from a required or 'foundational' field to an 'elective' subject. This was interpreted by many academics as a disciplinary demotion and challenge to the institutionalization of a field, which, in BHU, draws its authority primarily from governmental approval and student demand.

12 Some instructors supervise small groups of students as part of their *nauchno-issledovatel'skaia rabota so studentami* (scientific-research work with students). In 1996–97, Ibraeva mentored a group on 'the role of mass media information in the reformation of society' and Asanbekov supervised one which dealt with the 'problems of establishing new social commonalities' (BHU 1997). In 1999, Shaidullaeva organized a student club called 'Datkaiym' in order to hold discussions on contemporary social problems, particularly regarding female elites (BHU 2000).

13 From 1994–2000, the 'all-faculty scientific problem' was 'Kyrgyzstan on the road to democracy and the market'. By 1995, the department had developed its own 'all-department theme' within this broad framework, called 'Social changes in the conditions of a transitional society' (BHU 1995, 1998, 2002).

14 Under the first theme, supervised by Isaev, staff conducted empirical research on topics such as the development of a national working class, entrepreneurs and farmers, the social problems of women, and the participation of young people in privatization, publishing twenty-one articles on the results of this research in national newspapers and several locally produced *sborniki* [collections of articles] (BHU 1995). The following year, the team developed a programme on 'monitoring public opinion', upon which basis they made recommendations to the governmental groups in charge of designing privatization policies and produced a four-part publication, *The Kyrgyz Republic: Changes in the Process of Social Transformation*, which focused on outlining the

effects of political and economic reform on everyday life in Kyrgyzstan (BHU 1995). The aim of this research was overwhelmingly to ascertain and expose the 'objective social reality' about the reforms, which, it was argued, was obscured by both popular misinterpretation and political propaganda.

15 The second theme, also supervised by Isaev, explored the 'formation of the political elite as it is directly linked with fundamental changes in the life of the new Kyrgyz state'. The process, it was asserted, could only occur in a democracy 'defined by political freedom and political pluralism'. Research on the topic, which was dominated by Isaev's controversial studies on political ratings conducted from 1991–97), therefore focused on drawing correlations between elite power and levels of political freedom in the republic (BHU 1995).

16 For more on the history of the university, see Reeves (2003) and Sharshekeeva (2001).

17 Elective course options included comparative and historical sociology, sociology of culture, sociology of sex and gender, political sociology, sociological perspectives of mass media, racial and ethnic relations, environmental sociology and human ecology, social demography, principles and methods of computerized statistics and collective behaviour and social movements. Draft documents from this period also list other potential course offerings such as comparative Marxism, criminology, conflict resolution, medical sociology, democracy and institutions, deviance and social control, sociology of education, social history and urban sociology.

18 Categorical distinctions between 'indigenous' and 'Western' forms of knowledge are epistemologically difficult to sustain from a philosophy of knowledge perspective, within which it makes 'more sense to talk about multiple domains and types of knowledge' which intersect under particular circumstances (Agrawal 1995). Further research is necessary to understand precisely how these terms are being constructed and used in Central Asia.

19 To overcome this and encourage instructors to be more creative, he lobbied the administration to make their courses compulsory (required courses are held regardless of student numbers). He and a number of others from the Sociology Department also initiated a faculty-staff union, which had strong beginnings but dissolved during the university's administrative crisis.

20 The most established is Ablezova and Botoeva's work on developing methodologies for large-scale empirical studies of social problems that are of interest to international aid agencies. While Sagynbaeva's early work on qualitative methodology, particularly the use of focus groups for marketing studies, was discontinued after she left the department in 2002, undergraduate students now gain practical experience with interviewing, focus groups and questionnaire construction through the Applied Research Center or local market research companies.

7 PUBLIC SOCIAL SCIENCE IN CENTRAL ASIA

1 For an alternative interpretation of sociology in the media, in particular its 'destructive effects' and efforts to regulate the publication of sociological work, see Fond Zashchity Glasnosti (1996).

2 The Sociology Department at BHU, for example, includes newspaper articles and reports in its annual research reports. The fact that such publications are not considered contributions towards the fulfilment of a candidate or doctoral degree, however, suggests that there is still considerable stigma attached to media publication.

3 Other papers in which articles on sociology have appeared include *Asaba, Ata Zhurt, Betme-bet, Bizinesmen Kyrgyzstana, Vestnik vremeni, Vechernii Bishkek, Delo No., Zaman, Zhurnalist, Kommunist Kirgizstana, Komsomolets Kirgizii, Kut bilim, Kyrgyz rukh, Leninskii put' (Osh), Liberal'naia gazeta, Liudi i svet, Molodezhnaia*

gazeta, Moskovskom Komsomol'tse, Mugalimder gazetasy, Nasha gazeta, Nauka i tekhnika, Osh zhanyryty, Pami, Panerama, Politika i obschestvo, Pravda, Propagandist i agitator Kirgizstan, Piatnitsa, ResPublica, Reforma, Rynok kapitalov, Svobodnye gory/Erkin too, Slovo Kyrgyzstana, Sovetskaia Kirgiziia, Stolitsa, Trudy Kirgizii, Utro Bishkeka, Uchitel' Kirgizstana, Iuzhnyi kur'er and *Ekho Osha.*

4 It is difficult to obtain reliable statistical data on the expansion of publications on public opinion research during the early independence period. Many individuals and organizations that have conducted and published public opinion surveys in Kyrgyzstan during the last decade were unregistered, and many groups or centres dissolve within months or years. There have been no regularly produced academic periodicals in the republic since the early 1990s and studies are often published in newspapers, but these articles often lack authorship or are printed under pseudonyms. Furthermore, many research centres do not keep accurate accounts of the studies they themselves conduct and those that do are often unwilling to share their archives with outsiders.

5 *Prikhvatizatsiia* is an ironic pun on 'privatization', stemming from the Russian verb *prikhvatit'*, or 'to seize up'. It translates loosely into crony capitalism and refers to the post-independence movement in which state land was redistributed to wealthy *apparatchiki* and oligarchs rather than genuinely privatized.

6 The articles are: Isaev (1994, 1994a,b); Isaev, Akbagysheva and Abylagazieva (1994); Isaev and Abylgazieva (1994); Isaev and Asanbekov (1994); and Osmonalieva (1994).

7 For more on Akaev's increasingly authoritarian governance, see Spector (2004).

8 For more on privatization in the Kyrgyz Republic, see Dabrowski *et al.* (1995) and Nichols (1997).

9 'National-level' sociological studies in Kyrgyzstan are not necessarily based on representative samples of the national population; in fact, many use localized samples and generalize them to the 'nation'. The practice of using proportional as opposed to representative samples in sociological research, and its justification as being superior for studies in the largely rural republic, will be discussed later in the chapter.

10 In one, for example, Isaev and Asanbekov (1994) argued that ethnically Kyrgyz respondents were least informed about privatization not only because they had access to fewer media sources in the rural regions of the republic, but also because they maintained a more 'traditional' way of life in which information is communicated through informal relationships as opposed to official networks such as the media. They also suggested, as is commonly argued in Kyrgyz ethnology, that the 'nomadic past' of the Kyrgyz people dominated their collective economic psychology (or 'mentality' in local terms) to such an extent that it prevented them from being independently minded, and that they thus would fare better under programmes for more communal forms of privatization.

11 This set of 20 articles is comprised of Bakir Uluu (1994, 1997); Bekturganov (1994, 1995); Isaev (1996); Isaev and Ibraeva (1996); Isaev, Ibraeva, and Madaliev (1995); Isaev *et al.* (1993a, 1994b,c,d,e,f,g, 1996a,b, 1997, 1997a); Razguliaev (1995); Zhorobekova *et al.* (1995).

12 It is unclear when exactly these surveys began; one source traces them back to 1991 (Isaev 1991b), another to 1992 (Isaev and Ibraeva 1995), and yet another to 1994 (Isaev *et al.* 1997). During 2003, political ratings were regularly published in the weekly newspaper *Obschestvennyi reiting* [*Social Rating*].

13 It is worth noting that *Propagandist i agitator Kirgizstana* [*Propagandist and Agitator of Kirgizstan*] regularly published articles about sociology or written by sociologists during the 1980s.

14 Isaev was at the time a member of the Democratic Movement of Kyrgyzstan (DDK), which was founded in 1990 and 'served as an umbrella for a number of pro-democracy

and nationalist groups.... It backed the election of Akaev to the presidency in 1991, but later withdrew its support'. Bakir Uluu was a member of the Democratic Party of Free Kyrgyzstan (Erkin or ERK), which was 'founded in 1991 as a splinter group of the DDK on a platform of moderate nationalism and support a liberal market economy' (from Swiss Agency 2000).

15 Quote excerpted from Bakir Uluu (1994).

16 While Merton's (1942) 'ethos of science' is often accepted as a descriptive model of the culture of Western science, sociologist of science Mulkay (1976, 1979) reinterprets it as a normative prescription for scientific practice, one that is tied in with the professional and societal conditions of modern science. See also Gieryn (1983: 783).

WORKS CITED

Abazov, R. (1989) 'Natsional'naia politika: puti resheniia' [National politics: paths to resolution], *Komsomolets Kirgizii*, 22 November: 8.

Abdrakhmanova, A. (2004) 'Kyrgyzstan: top university rocked by dispute', *Institute for War and Peace Reporting* (RCA No. 267, 20 February 2004). Available online at: www.iwpr.net/index.pl?archive/rca/rca_200402_267_2_eng.txt (accessed 5 April 2005).

Abdurakhimova, N. (2002) 'The colonial system of power in Turkistan', *International Journal of Middle East Studies*, 34: 239–62.

Abdyrashev, M. (1994) 'Sotsiologiia do vostrebovaniia' [Sociology *poste restante*], *Slovo Kyrgyzstana*, 4 January.

Ablezova, M. (2003) Interview by the author, American University–Central Asia, Bishkek, Kyrgyzstan, 20 March.

Adams, L. (1999) 'The mascot researcher: identity, power and knowledge in fieldwork', *Journal of Comparative Ethnography*, 28(4): 331–63.

Admaliev, K. and Tuzov, A. (1991) 'Otkrytye otvety na "otkrytye voprosy"' ['Open answers to "open questions"'], *Vechernyi Bishkek*, 5 July.

Agrawal, A. (1995) 'Dismantling the divide between indigenous and western knowledge', *Development and Change*, 26(3): 413–39.

Aitmatov, C. (1983) *The Day Lasts More than a Hundred Years*, Bloomington and Indianapolis, IN: Indiana University Press.

Akaev, A. (1991) 'Address by Askar A. Akaev, President of the Republic of Kirgizstan, at the plenary meeting of the forty-sixth session of the United Nations general assembly – 22 October', in Furtado, C. and Chandler, A. (eds) *Perestroika in the Soviet Republics: Documents on the national question*, Boulder, CO: Westview Press.

Ake, C. (1982) *Social Science as Imperialism: The theory of political development*, Nigeria: Ibadan University Press.

Akiwowo, A. (1999) 'Indigenous sociologies: extending the scope of the Argument', *International Sociology*, 14(2): 115–38.

Alatas, S. (2000) 'Intellectual imperialism: definition, traits and problems', *Southeast Asian Journal of Social Sciences*, 28(1): 23–45.

—— (2003) 'Academic dependency and the global division of labor in the social sciences', *Current Sociology*, 51(6): 599–613.

Aldasheva, A. (2003) Interview by the author, Bishkek Humanitarian University, Bishkek, Kyrgyzstan, 22 May.

Aldasheva, Sh. and Nikolaenko, K. (1973) 'O nekotorykh rezul'tatakh sotsiologichestkikh issledovanii na predpriatiakh goroda Dzhala-Abada' [On some results of sociological research in the industrial enterprises of the city of Jalal-Abad], *Informatsionnyi listok*, 84: 1.

Alexander, J. and Colomy, P. (1992) 'Tradition and competition: preface to a postpositivist approach to knowledge cumulation', in Ritzer, G. (ed.) *Metatheorizing*, London: Sage Publications.

Ali, A. S. (1964) *The Modernization of Soviet Central Asia*, Lahore: Punjab University.

Alimova, B. (1984) 'O podgotovke nauchnykh-obschestvovedov Kirgizii v period razvitogo sotsializma' [On the preparation of scientific social scientists of Kirgizia in the period of developed socialism], *Molodye obschestvovedy – 60-letnemu iubeleiu sovetskogo Kirgizstana* [Young social scientists – the sixty year jubilee of Soviet Kyrgyzstan], Frunze: Academy of Science of the Kirgiz SSR, Institute of History.

Allworth, E. (1975) *Soviet Asia: Bibliographies – a compilation of social science and humanities sources on the Iranian, Mongolian and Turkic Nationalities, with an eessay on the Soviet–Asian controversy*, New York: Praeger Publishers.

—— (1998) 'History and group identity in Central Asia', in Smith, G., Law, V., Wilson, A., Bohr, A. and Allworth, E. (eds) *Nation Building in the Post-Soviet Borderlands: The politics of national identity*, Cambridge: Cambridge University Press.

Altbach, P. (1971) 'Education and neo-colonialism: a note', *Comparative Education Review*, 15(2): 237–39.

American University in Kyrgyzstan (2002) Undergraduate course catalog. Bishkek: American University in Kyrgyzstan.

American University in Kyrgyzstan Department of International Relations (1998) Draft of undergraduate programme curriculum, unpublished document, departmental archive.

—— (1998a) Sociology Department annual plan, unpublished document, departmental archive.

—— (1999) Promotional brochure, unpublished document, departmental archive.

—— (1999a) 'Prioritetnye napravleniia razvitiia Programmemy Sotsiologiia' [Priority directions for the development of the Sociology Programme], unpublished document, departmental archive.

—— (1999b) Needs assessment for the academic year 1999–2000, unpublished document, departmental archive.

—— (1999c) Professor–instructor roster, unpublished document, departmental archive.

—— (2000) 'Nagruzka prepodavatelei napravleniia "Sotsiologiia" na I semester 2000–01 uch. god' [List of instructors in the sociology programme for the first semester of the 2000–01 academic year], unpublished document, departmental archive.

—— (2000a) 'Svedeniia o kadrovom obespechenii obrazovatel'nogo protsessa (sovmestiteli) Amerikanskii Universitete v Kyrgyzstane' [Information about cadre provisions for the educational process (affiliate faculty) – American University in Kyrgyzstan], unpublished document, departmental archive.

—— (2001) Proposed budged for the academic year 2001–2, unpublished document, departmental archive.

—— (2002) Curriculum structure, second draft, unpublished document, departmental archive.

—— (2003) AUCA Applied Research Center, unpublished document, departmental archive.

—— (2003a) Sociology Department needs assessment, unpublished document, departmental archive.

—— (2003b) Departmental brochure, first draft, unpublished document, departmental archive.

—— (2003c) Departmental brochure, second draft, unpublished document, departmental archive.

—— (2003d) Notes from joint Sociology and International Comparative Politics faculty meeting, unpublished document, departmental archive, 5 June.

'Appeal by the Central Committee of the Communist Party of Kirgizia, the Presidium of the Supreme Soviet, and the Council of Ministers of the Kirgiz Soviet Socialist Republic – 9 June' (1992) in Furtado, C. and Chandler, A. (eds) *Perestroika in the Soviet Republics: Documents on the national question*, Boulder, CO: Westview Press.

'Appeal from the USSR Supreme Soviet to the People of the Kirgiz SSR on the Violence in Osh – 8 June (1990)', in Furtado, C. and Chandler, A. (eds) *Perestroika in the Soviet Republics: Documents on the national question*, Boulder, CO: Westview Press.

Asanbek Tabaldiev (1975) 'Obituary', in *Sovietskaia Kirgiziia*, 16 December: 6.

Asanbekov, M. (2003) Interview by the author, Bishkek Humanitarian University, Bishkek, Kyrgyzstan, 6 February.

Asanova, U. (1995) 'Nuzhno prizvat' vse muzhestvo intellektual'noi sovesti' [We must gather all our intellectual courage], *Kut Bilim*, 15 February: 6.

—— (2003) Interview by the author, American University–Central Asia, Bishkek, Kyrgyzstan, 4 June.

Bagramov, E. (1990) 'The national problem and social science', *Pravda*, 16 August: 2–3; reprinted in Olcott, M., Hajda, L. and Olcott, A. (eds) *The Soviet Multinational State: Readings and documents*, New York: M.E. Sharpe.

Baibosunov, K. (1993) 'Dama bez kaprizov, ili kak my ponimaem nauku sotsiologii' [A lady without whims, or how we understand the science of sociology], *ResPublica*, 29 October.

—— (1998) Interview with K. Isaev, 'Azyrky Kyrgyz sotsiologiasynyn maseleleri' [Modern sociological issues on the agenda in Kyrgyzstan], *Kyrgyz Rukhu*, 30 April.

Bailey, L. (1996) *Critical Theory and the Sociology of Knowledge: A comparative study in the theory of ideology*, New York: Peter Lang.

Bakir Uluu, T. (1994) 'Zerkal'noe bolezn' "karavanshika sotsiologii" Kuseina Isaeva' [The reflective illness of Kusein Isaev's 'Caravan Sociology'], *ResPublica*, 5 February.

—— (1997) 'V chem ia provnil'sia narodam?' [In what way have I wronged the people?], *Slovo Kyrgyzstana*, 17–18 January: 5.

Balázs, K., Faulkner, W. and Schimank, U. (1995) 'Transformation of the research systems of post-communist Central and Eastern Europe: an introduction', *Social Studies of Science*, 25(4): 613–32.

Bartlett, P. (2006) 'Speak no English', Transitions. Available online at: www.tol.cz (accessed 25 August 2006).

Batygin, G. and Deviatko, I. (1994) 'The metamorphoses of Russian sociology', in Keen, M. F. and Mucha, J. (eds) *Eastern Europe in Transition: The impact on Sociology*, Westport, CT: Greenwood Press.

Beck, U. (2000) 'The cosmopolitan perspective: sociology for the second age of modernity', *British Journal of Sociology*, 51: 79–106.

—— (2002) 'The cosmopolitan society and its enemies', *Theory, Culture & Society*, 19(1–2): 17–44.

Beissinger, M. (1988) *Scientific Management, Socialist Discipline and Soviet Power*, London: I. B. Taurus.

WORKS CITED

Beissinger, M. and Young, C. (2002) *Beyond State Crisis? Postcolonial Africa and Post-Soviet Eurasia in Comparative Perspective*, Washington, DC: Woodrow Wilson Center Press.

Bekturganov, K. (1990) 'Obschestvennoe mnenie i politiki' [Public opinion and politics], *Kommunist Kirgizstana*, 8/9: 106.

Bekturganov, K. (1991) 'K voprosu sotsiologicheskogo izucheniia sostoianiia internatsional'nogo vospitaniia molodezhi' [Toward the sociological study of the state of the internationalisation of raising young people], *Problems of Raising Young People: A collection of scientific works*, Bishkek.

—— (1994) 'Sotsiologiia obschestvennogo mneniia: metodologicheskie i organizatsionno–metodoicheckie aspeckty' [The sociology of public opinion: methodological and organizational–methodic aspects], *Vestnik KGNU*, Social Science Series.

—— (1994a) 'Professionaly ne molchat, a bliudut chest' mundira' [Professionals do not remain silent, but guard the honour of the uniform], *Slovo Kyrgyzstana*, 3 December.

—— (1995) 'Portret prezidenta nashimi glazami' [A portrait of the president with our eyes], *Slovo Kyrgyzstana*, 9–10 December: 7.

—— (1995a) 'Problemy transformatsii sotsiologicheskogo obrazovaniia v novykh usloviakh' [Problems in the transformation of sociological education in the new conditions], *Education and Science in the New Geopolitical Space*, Scientific-practical conference, Bishkek.

—— (1997) 'Kyrgyzstan na pereput': sotsiologicheskii srez' [Kyrgyzstan at the crossroads: a sociological cross–section], *Kut Bilim*, 31 May and 5 June.

—— (2003) Interview by the author, Kyrgyz National University, Bishkek, Kyrgyzstan, 25 June.

Bekturganov, K., Omurbekov, T. and Tishin, A. (1994) 'Komu vygoden populism v sotsiologii' [Who benefits from populism in sociology?], *Slovo Kyrgyzstana*, 11 May.

Beliaev, E. and Butorin, P. (1982) 'The institutionalization of Soviet sociology: its social and political context', *Social Forces*, 61(2): 418–35.

Bess, J. (2000) 'The dilemmas of change: higher education in Belarus', *International Higher Education*, Winter.

Bishkek Humanitarian University, Sociology Department (1994) Plan raboty kafedry sotsiologii i politologii na 1994–95 uchebnom godu, BHU [Annual plan of the Sociology Department and Politology for the 1994–95 academic year], unpublished document, Bishkek Humanitarian University archive.

—— (1995) Otchet o nauchno–issledovatel'skoi rabote kafedry sotsiologii i politologii za 1995 god [Report on scientific research work in the Sociology Department and Politology during 1995], unpublished document, Bishkek Humanitarian University archive.

—— (1995a) Plan raboty kafedry sotsiologii i politologii na 1995–6 uchebnom godu, BHU [Annual plan of the Sociology Department and Politology for the 1995–6 academic year], unpublished document, Bishkek Humanitarian University archive.

—— (1996) Plan raboty kafedry sotsiologii i politologii na 1996–7 uchebnom godu, BHU [Annual plan of the Sociology Department and Politology for the 1996–7 academic year], unpublished document, Bishkek Humanitarian University archive.

—— (1997) Otchet rabote kafedry sotsiologii za 1996–7 uch. god [Report of work in the Sociology Department for the 19967 academic year], unpublished document, Bishkek Humanitarian University archive.

—— (1997a) Plan raboty kafedry sotsiologii i politologii na 1997–8 uchebnom godu, BHU [Annual plan of the Sociology Department and Politology for the 1997–8 academic year], unpublished document, Bishkek Humanitarian University archive.

—— (1997b) Protokol No. 12, Zasedanie kafedry sotsiologii ot 02.07.97 [Protocol No. 12: sociology department meeting of 7 February 1997], unpublished document, Bishkek Humanitarian University archive.

—— (1998) Otchet o nauchno–issledovatel'skoi rabote kafedry sotsiologii i politologii za 1998 god [Report on scientific research work in the Sociology Department and Politology during 1998], unpublished document, Bishkek Humanitarian University archive.

—— (2000) Otchet rabote kafedry sotsiologii za 1999–2000 uch. god [Report of work in the Sociology Department for the 1999–2000 academic year], unpublished document, Bishkek Humanitarian University archive.

—— (2001) Protokol No. 5, Zasedanie kafedry sotsiologii ot 16.01.01 [Protocol No. 5: sociology department meeting of 16 January 1997], unpublished minutes of faculty meeting, unpublished document, Bishkek Humanitarian University archive.

—— (2002) Otchet o nauchno-issledovatel'skoi rabote kafedry sotsiologii i politologii za 2001–2 god [Report on scientific research work in the Sociology Department and Politology for 2001–2], unpublished document, Bishkek Humanitarian University archive.

—— (2002a) Zasedanie kafedry sotsiologii ot 12.02.02 [Sociology department meeting of 12 February 2002], unpublished document, Bishkek Humanitarian University archive.

—— (2002b) Protokol, zasedanie kafedry sotsiologii ot 18.08.02 [Protocol, sociology department meeting of 18 August 2002], unpublished document, Bishkek Humanitarian University archive.

—— (2003) Promotional brochure, unpublished document, Bishkek Humanitarian University archive.

—— (2003a) Protokol, zasedanie kafedry sotsiologii ot 29.01.03 [Protocol, sociology department meeting of 29 January 2003], unpublished document, Bishkek Humanitarian University archive.

Bitkovskaia, G. (1996) 'Migranty stremiatsia v Rossiiu' [Migrants strive for Russia], *Slovo Kyrgyzstana*, 19–20 July.

Bloor, D. (1991) *Knowledge and Social Imagery*, London: Routledge.

—— (1999) 'Anti-Latour', *Studies in History and Philosophy of Science*, 30(1): 81–112.

Blum, Y. (1990) 'Professora i professionally: zametki s pervoi respublikanskoi konferentsii sotsiologov' [Professors and professionals: remarks from the first republican conference of sociologists], *Sovietskaia Kirgiziia*, 27 January: 3.

—— (1991) 'Prestizh professi v natsional'nom razreze' [The prestige of professions in national cross–section], *Slovo Kyrgyzstana*, 23 November.

—— (1993) 'Pravda – ne v sile, no sile – v pravde' [Truth is not in strength, but strength is in truth], *Slovo Kyrgyzstana*, 6 February: 7.

Blume, S. (1974) *Toward a Political Sociology of Science*, New York: The Free Press.

Boronoev, A. (1999) 'Fakul'tet sotsiologii: 10 let stanovleniia i razvitiia' [The faculty of sociology: 10 years of establishment and development], *Zhurnal sotsiologii i sotsial'noi antropologii,* 2(2). Available online at: www.soc.pu.ru:8101/publications/jssa/1999/3/8grig.html (accessed 17 November 2004).

Borozdin, N. (1929) 'Inter-racial study in Asia', *Pacific Affairs*, 2(6): 322–28.

Botoeva, G. (2003) 'Interview by the author', American University–Central Asia, Bishkek, Kyrgyzstan, 26 May.

Bourdieu, P. (1975) 'The specificity of the scientific field and the social conditions of the progress of reason', *Social Science Information*, 14(6): 19–47.

—— (1988) *Homo Academicus*, Cambridge: Polity Press.

Bourdieu, P. (1991) 'The peculiar history of scientific reason', *Sociological Forum*, 6(1): 3–26.

—— (1992) *The Logic of Practice*, Cambridge: Polity Press.

—— (1993) *The Field of Cultural Production*, Cambridge: Polity Press.

—— (1999) 'The space of points of view', in *The Weight of the World: Social suffering in contemporary society*, Cambridge: Polity Press.

Bronson, S., Popson, N. and Ruble, B. (1999) 'Sustaining intellectual communities: a strategy for rebuilding the social sciences and humanities in the former Soviet Union from within', in *The Humanities and Social Sciences in the Former Soviet Union: An assessment of need*, unpublished report prepared by the Kennan Institute, Woodrow Wilson Center for the Carnegie Corporation of New York and the MacArthur Foundation.

Bruno, M. (1998) 'Playing the cooperation game: strategies around international aid in post-socialist Russia', in Pine, F. and Bridger, S. (eds) *Surviving Post-socialism: Local strategies and regional responses in eastern Europe and the former Soviet Union*, London and New York: Routledge.

Bryant, C. and Mokrzycki, E. (eds) (1994) *The New Great Transformation? Change and Continuity in East-Central Europe*, London: Routledge.

Brym, R. (1990) 'Sociology, perestroika, and Soviet society', *The Canadian Journal of Sociology*, 15(2): 207–15.

Buckley, C. (1999) 'Ideology, methodology, and context: social science surveys in the Russian Federation', *American Behavioral Scientist*, 43(2): 222–36.

Bujra, A. (1994) 'Whither social science institutions in Africa: a prognosis', *Afrique et Develeppement* [Africa and Development], Special Issue on the Social Sciences in Post–Independent Africa: Past, Present, and Future, 19(1): 119–66.

Burawoy, M. (1992) 'The end of Sovietology and the renaissance of Modernization Theory', *Contemporary Sociology*, 21(6): 774–85.

—— (2000) 'Reaching for the global', in *Global Ethnography: Forces, connections, and imaginations in a postmodern world*, Berkeley, CA: University of California Press.

Butenko, I. A. (2002) 'The Russian Sociological Association: actors and scenery on a revolving stage', *International Sociology*, 17(2): 233–51.

Camic, C. (1995) 'Three departments in search of a discipline: localism and interdisciplinary interaction in American Sociology, 1890–1940', *Social Research*, 62: 1003–33.

Camic, C. and Xie, Y. (1994) 'The statistical turn in American social science: Columbia University, 1890–1915', *American Sociological Review*, 59(October): 773–805.

Caroe, O. (1953) 'Soviet colonialism in Central Asia', *Foreign Affairs*, 32(1): 134–44.

Cavenaugh, C. (2001) 'Central Asia's colonial past and why it matters', *Central Asia Monitor*, 5/6: 9–17.

Chekki, D. (1987) *American Sociological Hegemony: Transnational explorations*, New York: University Press of America.

Chomsky, N. (2000) *Chomsky on MisEducation*, Oxford: Rowman and Littlefield.

Chomsky, N., Nader, L., Wallerstein, I., Lewontin, R., Ohmann, R., Zinn, H., Katznelson, I., Montgomery, D. and Siever, R. (eds) (1997) *The Cold War and the University: Toward an intellectual history of the postwar years*, New York: New Press.

Chrissterson, L. (1994) 'Soviet and post-Soviet science', *Science*, 265(5170): 301.

Civic Education Project (2000) 'Strategic plan for Civic Education Project (CEP), July 2000 to June 2003', unpublished document.

—— (2003) 'Programme strategy, 2004–06', unpublished document.

Clem, R. (1992) 'The frontier and colonialism in Russian and Soviet Central Asia', in Lewis, R. (ed.) *Geographic Perspectives on Central Asia*, London and New York: Routledge.

Clinard, M. B. and Elder, J. W. (1967) 'Sociology in India: a study in the sociology of knowledge', *American Sociological Review*, 30(August): 581–87.

Cole, S. (2001) *What's Wrong with Sociology?* Somerset, NJ: Transaction Publishers.

Collins, H. and Pinch, T. (1993) *The Golem: What everyone should know about science*, Cambridge: Cambridge University Press.

Comte, A. ([1853] 1975) in Lenzer, G. (ed.) *August Comte and Positivism: The essential writings*, New York: Harper and Row.

Cooper, F. and Packard, R. (1997) *International Development and the Social Sciences: Essays on the history and politics of knowledge*, Berkeley, CA: University of California Press.

Council of Europe (2000) *Social Sciences and the Challenge of Transition*. Available online at: http://culture.coe.fr (accessed 30 May 2005).

Cozzens, S. and Gieryn, T. (1990) *Theories of Science in Society*, Bloomington, IN: Indiana University Press.

Critchlow, J. (1972) 'Signs of emerging nationalism in the Moslem Soviet Republics', in Dodge, N. (ed.) *The Soviets in Asia*, proceedings of a symposium sponsored by the Washington Chapter of the American Association for the Advancement of Slavic Studies and the Institute for Sino–Soviet Studies, George Washington University, 19–20 May, Washington, DC: Cremona Fund.

Dabrowski, M., Jermakowicz, W., Pańków, J., Kloc, K. and Antczak, R. (1995) 'Economic reforms in Kyrgyzstan', *Communist Economies and Economic Transformation*, 7(3): 269–97.

De Young, A. (2001) 'On the prospects for secondary education reform in the Kyrgyz Republic', *Central Asia Monitor*, (4): 5–15; (5-6): 46–58.

'Dlia nauki net granits' [No borders for science] (1999) *ResPublica*, July [no date].

Dudwick, N. and De Soto, H. (2000) 'Introduction', in De Soto, H. and Dudwich, N. (eds) *Fieldwork Dilemmas: Anthropologists in postsocialist states*, Wisconsin: University of Wisconsin Press.

Dunston, J. (ed.) (1992) *Soviet Education under Perestroika*, New York: Routledge.

Durkheim, E. (1938) *The Rules of Sociological Method*, New York: Free Press.

Dzhangirov, E., Khorolets, S. and Isaev, K. (1987) 'Otchet o nauchno-issledovatel'skoi rabote: razrabotka rekomendatsii k planu sotsial'nogo razvitiia sovkhoza "Stavropol'skii" na XII piatiletku i na period do 2000 goda (zakliuchitel'nyi)' [Report on scientific research: the development of recommendations toward a social development plan for the 'Stavropol'skii' state farm during the 12th five-year plan and in the period until the year 2000], Frunze: Ministry of Education and Special Middle Education of the Kirgiz SSR and the Frunze Polytechnic Institute.

Eades, J. and Schwaller, C. (eds) (1991) *Transitional Agendas: Working papers from the summer school for Soviet sociologists*, Centre for Social Anthropology and Computing, Canterbury: University of Kent.

Editor (1998) 'Obraschenie k chitateliam Uzbekistanskogo gumanitarnoho zhurnala *Obschestvennoe mnenie*' [To the readers of *Public Opinion*, Uzbekistan's Humanitarian Journal], *Obschestvennoe mnenie*, 1: 9.

Egorov, I. (2002) 'Perspectives on the scientific systems of the post–Soviet states: a pessimistic view', *Prometheus*, 20(1): 59–73.

Eisemon T. (1982) *The Science Profession in the Third World: Studies from India and Kenya*, New York: Praeger.

Eisenstadt, S. (1992) 'Centre–periphery relations in the Soviet empire: some interpretive observations', in Motyl, A. (ed.) *Thinking Theoretically about Soviet Nationalities: History and comparison in the study of the USSR*, New York: Columbia University Press.

Elebaeva, A., Dzhusupbekov, A. and Omuraliev, N. (1991) *Oshskii mezhnatsional'nyi konflikt: sotsiologicheskii analiz* [The Osh international conflict: a sociological analysis], Centre for Social Research, National Academy of Sciences of Kyrgyzstan.

—— (1995) *Razvitie mezhnatsional'nykh otnoshenii v novykh gosudarstvakh Tstentral'noi Azii* [The development of international relations in the new states of Central Asia], Bishkek: Ilim.

Escobar, A. (1991) 'Anthropology and the development encounter: The making and marketing of development anthropology', *American Ethnologist*, 18(4): 658–81.

Fanisov, R. (1990) 'Stho mozhet sotsiologiia?' [What can sociology do?], *Uchitel' Kyrgyzstana*, 6 April: 7.

Fanon, F. (1963) *The Wretched of the Earth*, London: Penguin.

Fernandes, F. (1967) 'The social sciences in Latin America', in Diéges Júnior, M. and Wood, B. (eds) *Social Science in Latin America*, papers presented at the conference on Latin American Studies held at Rio de Janeiro, 29–31 March 1965, New York: Columbia University Press.

Feyerabend, P. (1981) *Realism, Rationalism and Scientific Method*, Cambridge: Cambridge University Press.

—— (1993) *Against Method*, 3rd edn, London and New York: Verso.

Fierman, W. (1990) Review of Rumer, B. Z., 'Soviet Central Asia: a tragic experiment', in *The American Political Science Review*, 84(3): 1047–48.

Filino, E. D. L. (1990) 'Sociology and society in Brazil and Argentina 1954–1985', PhD dissertation, Brown University.

Fisher, D. (1990) 'Boundary-work and science: the relationship between knowledge and power', in Cozzens, S. and Gieryn, T. (eds) *Theories of Science in Society*, Bloomington, IN: Indiana University Press.

Fond Zashchity Glasnosti (1996) *Sotsiologiia i pressa v period parliamentskix i presidentskix vyborov 1995 i 1996 godov* [Sociology and the press in the period of parliamentary and presidential elections of 1995 and 1996], Moscow: Izdatel'stvo Prava Cheloveka.

Foucault, M. (1967) *Madness and Civilisation: A history of insanity in the age of reason*, London: Tavistock.

—— (1973) *The Birth of the Clinic: An archaeology of medical perception*, London: Tavistock.

—— (1989) *The Archaeology of Knowledge*, London: Routledge.

—— (2001) *The Order of Things*, London: Routledge.

Fuchs, S. (1986) 'The social organization of scientific knowledge', *Sociological Theory*, 4(Fall): 126–42.

Fuller, S. (1993) 'Disciplinary boundaries and the rhetoric of the social sciences', in Messer-Davidow, E., Shumway, D. and Sylvan, D. (eds) *Knowledges: Historical and critical studies in disciplinarity*, Charlottesville, VA and London: University Press of Virginia.

Galtung, J. (1971) 'A structural theory of imperialism', *Journal of Peace Research*, 2: 81–117.

Gammer, M. (2000) 'Post-Soviet Central Asia and postcolonial Francopone Africa: some associations', *Middle Eastern Studies*, 36(2): 124–49.

Ganon, I. (1965) 'What is national sociology?', *Sociologia*, 5(8–9): 9–23.

Gaponenko, N. (1995) 'Transformation of the research system in a transitional society: the case of Russia', *Social Studies of Science*, 25(4): 685–703.

Garreau, F. (1988) 'Another type of third world dependency: the social sciences', *International Sociology*, 3(2): 171–78.

Gasper, P. (1998) 'Bookwatch: Marxism and science', *International Socialism*, 79: 137–71.

Gellner, E. (1988) *State and Society in Soviet Thought*, Oxford: Basil Blackwell.

Gendzier, I. (1985) *Managing Political Change: Social scientists and the third world*, Boulder, CO: Westview Press.

Genov, N. (1989) 'National traditions in Sociology', *Sage Studies in International Sociology*, 36, London: Sage.

Gieryn, T. (1983) 'Boundary-work and the demarcation of science from non-science: strains and interests in professional ideologies of scientists', *American Sociological Review*, 48(December): 781–95.

Gieryn, T., Bevins, G. and Zehr, S. (1985) 'Professionalization of American scientists: public science in the creation/evolution trials', *American Sociological Review*, 50(June): 392–409.

Gleason, G. and Buck, S. (1993) 'Decolonization in former Soviet borderlands: politics in search of principles', *Political Science and Politics*, 26(4): 522–25.

Glenady, C. (1996) 'The Republic of Kyrgyzstan: an overview of science, technology and higher education', Report 77, April–May. Available online at: www.nsf.gov/pubs/stis1996/int9612/int9612.txt (accessed 19 December 2004).

Goldfarb, J. (1990) 'Review of Zaslavskaia, T.', *A Voice of Reform in Contemporary Sociology*, 19(1): 107–09.

Good, J. (2000) 'Disciplining social psychology: a case of boundary relations in the history of the human sciences', *Journal of the History of the Behavioral Sciences*, 36(4): 383–403.

Gosovic, B. (2000) 'Global intellectual hegemony and the international development agenda', *International Social Science Journal*, 166: 447–56.

Graham, L. (1998) *What Have We Learned about Science and Technology from the Russian Experience?* Stanford, CA: Stanford University Press.

Greenfeld, L. (1988) 'Soviet sociology and sociology in the Soviet Union', *Annual Review of Sociology*, 14: 99–123.

Grigorev, S. (1999) 'O roli fakul'tetov sotsiologii Sankt–Peterburgskogo (Leningradskogo) i Moskovskogo universitetov v stanovlenii i razvitie kompleksa "Sotsiologiia, psikhologiia i sotsial'naia rabota" AGU v 1980–99e gg' [The role of the St. Petersburg (Leningrad) and Moscow faculties of sociology in the creation and development of the complex 'Sociology, psychology and social work' at the Altai State University from 1980–99], *Zhurnal sotsiologii i sotsial'noi antropologii*, 2(3). Available online at: www.soc.pu.ru:8101/publications/jssa/1999/3/8grig.html (accessed 7 February 2005).

Group of Independent Sociologists (1993) 'Kto pogodu delaet?' [Who controls the weather?], *Slovo Kyrgyzstana*, 1 May.

Gumport, P. and Snydman, S. (2002) 'The formal organization of knowledge: an analysis of academic structure', *Journal of Higher Education*, 73(3): 375–408.

Hahn, J. (1977) *Social Scientists and Policy Making in the USSR*, New York: Praeger.

Harris, R. (ed.) (1997) *Landmark Essays on Rhetoric of Science: Case studies*, Mahwah, NJ: Lawrence Erlbaum Associates.

Henderson, K. and Robinson, N. (1997) *Post-Communist Politics: An introduction*, Hertfordshire: Prentice Hall.

Hiller, H. (1979) 'Universality of science and the question of national sociologies', *The American Sociologist*, 14: 124–35.

Hollander, P. (1978) Comment on J. Kolaja, 'An observation on Soviet sociology', *Current Anthropology*, 19(2): 375–76.

Horkheimer, M. (1990 [1930]) 'A new concept of ideology?' in Meja, V. and Stehr, N. (eds) *Knowledge and Politics: The sociology of knowledge dispute*, London: Routledge, pp. 140–57.

Hulme, D. and Turner, M. (1990) *Sociology and Development: Theories, policies and practices*, New York: St. Martin's Press.

Huskey, E. (2004) 'From higher party schools to academies of state service: the marketization of bureaucratic training in Russia', *Slavic Review*, 63(2): 325–48.

Ibraeva, G. (2003) Interview by the author, written responses via email, February.

Inkeles, A. (1958) *Public Opinion in Soviet Russia: A study in mass persuasion*, Cambridge, MA: Harvard University Press.

Isaev, K. (1989) 'Kak pobedit' posredstvennost'?' [How can mediocrity be overcome?], *Sovietskaia Kirgiziia*, 25 July.

—— (1991)'Razdeliaiu tu mysl', shto tol'ko korova ne meniaet svoikh ubezhdenii' [I am of the opinion that only a sea–cow does not change its convictions], *Sovietskaia Kirgiziia*, 12 January.

—— (1991a) 'Sotsiologiia – i grazhdanskaia smelost', i naychaia dokazatel'nost'' [Sociology: both civic courage and scientific proof], *Komsomolets Kirgizie*, 13 February.

—— (1991b) 'Sotsiologiia v Kyrgyzstane: sostoianie i perspektivi' [Sociology in Kyrgyzstan: its status and perspectives], *Obshchestvennye nauki, Akademia nauki Respublikoi Kyrgyzstana*, 2: 27–34.

—— (1993) 'O meste i roli sotsiologii v sisteme sotsial'nykh predmetov vuzov Kyrgyzskoi Respubliki' [On the place and role of sociology in the system of social subjection in the higher education institutions of the Kyrgyz Republic], in *Kut Bilim*, 28 May.

—— (1993a) 'Kak sozdavalas' Kyrgyzskaia sotsiologicheskaia sluzhba' [How the Kyrgyz sociological service was created], *ResPublica*, 29 May.

—— (1993b) 'Narod vsegda prav, dazhe esli oshibaetsia....' [The people are always right, even when they are mistaken (rating of the leadership: results from the June survey)], *Slovo Kyrgyzstana*, 24 July: 6.

—— (1993c) 'Vozmozhno li Kyrgyzskaia tsivilizatsiia?' [Is a Kyrgyz civilisation possible?], *ResPublica*, 29 October.

—— (1994) 'Privatizatsiia: zachem? dlia kogo?' [Privatization: for what, for whom?], *Slovo Kyrgyzstana*, 21 January: 3.

—— (1994a) 'Narod i privatizatsiia: shto izmenilos'?' [The people and privatization: what changed?], *Slovo Kyrgyzstana*, 19 May.

—— (1994b) 'U kriticheskoi cherty' [From the critical line – sociological survey], *Iuzhnyi Kur'er*, 6, 12 February: 1.

—— (1995) 'Demokratsiia i vlast'' [Democracy and power], *Erkin-Too*, 30 September: 11.

—— (1996) 'Kto meshaet razvitie sotsiologii?' [Who interferes with the development of sociology?], *Erkin-Too*, 27 October.

—— (1996a) 'Moia svoboda, moia gordost'' [My freedom, my pride], *Vecherniy Bishkek*, 17 July. Also published in Isaev, K. 2003: 30–34.

—— (1997) 'Kyrgyzstane, Dzhordzh Soros, achyk koom' [Kyrgyzstan, George Soros, the open society], *Ata–Zhurt*, 8–18 May.

—— (1998) 'Sotsiologiiany saiasii soiku kylgylary bar' [They want to make a political prostitute out of sociology], *Asaba*, 6 March.

—— (1998a) 'Ochen' sovremennaia professiia' [A very modern profession], *RIF*, 26 June.

—— (1998b) 'Sotsiologiia Kyrgyzstana: v mire i u sebia' [Kyrgyz sociology: in the world and for itself], *Nasha Gazetta*, 26 September: 13.

—— (1998c) 'Byt duinodo sotsiologianyn kadyr barky osuudo' [The significance of the growth of sociology around the world], *Erkin-Too*, 14 December.

—— (1998d) 'Makhatma Gandi i otkrytoe obschestvo v Kyrgyzstane' [Mahatma Gandhi and the open society in Kyrgyzstan], *Slovo Kyrgyzstana*, 5 March.

—— (1999a) *Politicheskaia sotsiologiia: stanovlenie i razvitie* [Political sociology: its establishment and development], Bishkek.

—— (1999b) 'O radostiakh i ogorcheniiakh 1998 – go i nadezhdakh na 1999 – i' [On the celebrations and distresses of 1998 and hopes for 1999], *ResPublica*, 19–25 January.

—— (1999c) 'Mentalitet i obrazovannie' [Mentality and education], *Utro Bishkeka*, 24 November.

—— (1999d) 'Prezidentti "yiyk yiga" ailantpashybyz kerek' [Our president should not be considered a sacred cow], *Asaba*, 24 December.

—— (1999e) 'Orto mektepte sotsiologiia okulaby?' [Will sociology be taught at school?], *Mektep-shkola*, 4.

—— (2000) 'Problemy institutsionalizatsii nashei sotsiologii' [Problems of the institutionalisation of our sociology], *Utro Bishkeka*, 23–25 September: 15.

—— (2003) 'Istina stara kak mir' [A truth as old as the world], *Moskovskaia Komsomol'tse*, 22–29 March: 16.

—— (2003a) *Moia svoboda, moia gordost': posobie dlia izuchaiuschikh sotsiologiiu* [My freedom, my pride: a textbook for students of sociology], Bishkek.

Isaev, K. and Abylgazieva, A. (1994) 'Privatizatsiia: kazhdomu svoe' [Privatization: to each their own], source unknown.

Isaev, K. and Asanbekov, M. (1994) 'O chem shepshet vnutrennii golos' [What the inner voice whispers: ethnic particularities of privatization in Kyrgyzstan], *Slovo Kyrgyzstana*, no date.

Isaev, K. and Bekturganov, K. (1990). 'Obschestvennaia potrebnost' v sotsiologii' [Social demand in sociology], *Kommunist Kirgizstana*, 1: 3–8.

Isaev, K. and Ibraeva, G. (1995) 'O prezidentskikh vyborakh, o kachestve zhizni v proshlom i buduschem' [On the presidential election and the quality of life past and future], *Svobodnye gory*, 9 December: 6.

—— (1996) 'Politicheskii god Kyrgyzskoi elity v prostranstve obschestvennogo mneniia' [The political year of the Kyrgyz elite on the field of public opinion], *ResPublica*, 27 February.

Isaev, K. and Madaliev, M. (1998) 'Gde nakhodim'sia i kuda my idem?' [Where are we located and where are we going? (Analysis of a sociological survey)], *ResPublica*, 17–23 November.

Isaev, K. and Niyazov, E. (1990) 'Pervaia konferentsiia sotsiologov Kirgizii' [The first conference of sociologists of Kirgizia], *Sotsiologicheskie issledovaniia*, 7: 149–50.

Isaev, K., Niyazov, E. and Zhigitekov, K. (1993) 'Naroda vsegda prav, dazhe esli oshibaet-sia....' [The people are always right, even when they are mistaken], *Slovo Kyrgyzstana*, 24 July: 6.

Isaev, K., Niyazov, E. and Zhigitekov, K. (1993a) 'Sotsiologii i "besprovolochnyi telegraf"' Tursunbaia Bakir Uulu' [Sociology and the "wireless telegraph" of Tursunbai Bakir Uluu], *Slovo Kyrgyzstana*, 5 February: 7.

—— (1994) 'Sotsiologi staviat tochku nad 'I': starpom vesomei kapitana, ili Fenomen Apaca Zhumagulova' [Sociologists dot the *i*: measure of the captain's weight or the phenomenon of Apasa Zhmagulov], *Svobodnye gory*, 19 March: 1–2.

—— (1994a) 'Komy "piaterka" a komu "edinitsa": molodezh' o rukovoditeliakh respubliki' [Who gets a 'five' and who gets a 'one': youth on the leaders of the republic], *Slovo Kyrgyzstana*, 9 April.

—— (1994b) '50 politikov v Iiune' [Fifty politicians in June], *Slovo Kyrgyzstana*, 9 July: 10.

—— (1994c) 'Kogda tuz ne kozyr': politicheskaia situatsiia i reiting rukovoditelei' [When the ace isn't trump: the political situation and leaders' ratings], *Slovo Kyrgyzstana*, 17 September: 8.

—— (1994d) 'Politicheskie partii v sovremennoi sotsial'no–politicheskoi situatsii' [Political parties in the contemporary socio–political situation], *ResPublica*, 22 September.

—— (1994e) 'Komu otdat' svoi golos i nadezhdu' [To whom do you give your vote and hope?], *Slovo Kyrgyzstana*, 24 September: 5.

—— (1994f) '50 politikov Kyrgyzstana v sentiabre' [Fifty politicians of Kyrgyzstan in September], *ResPublica*, October [no date].

—— (1994g) 'Esli ne k khramu doroga, to zachem ona?' [If the road isn't to the church, what is it for? (The first president: results of his four years in perspective)], *Slovo Kyrgyzstana*, 15 November: 8.

Isaev, K., Akbagysheva, Z. and Abylagazieva, A. (1994) 'Minimum informatsii: tormoz privatizatsii?' [Is minimal information an obstacle to privatization?], *ResPublica*, 18 February.

Isaev, K., Ibraeva, G. and Madaliev, M. (1995) 'Noiabr'skie strasti olimpa' [The November passions of Olympus], *ResPublica*, 12 December.

Isaev, K., Akmatova, K. and Dosalieva, K. (1996) 'O proshlom i buduschem sotsial'nom samochuvstvii' [On the past and future of the general state of society], *ResPublica*, 24–29 April.

Isaev, K., Akmatova, K. and Sharshembieva, T. (1996) 'Tsennosti nashei zhizni' [Our life values], *Kut Bilim*, 24 October: 3.

Isaev, K., Madaliev, M. and Dosalieva, K. (1996) '40 vedyschikh politikov Kyrgyzstana v Aprele' [Forty well-known politicians of Kyrgyzstan in April], *ResPublica*, 5–10 June.

—— (1996a) 'Kto chego – to stoit – etogo ne skroet' [Someone values something – don't leave this out], *Slovo Kyrgyzstana*, 11–12 October.

—— (1996b) '40 vedyschikh politikov Kyrgyzstana v Noiabre' [Forty well–known politicians of Kyrgyzstan in November], *ResPublica*, 24–30 December.

—— (1997) '40 veduschikh politikov Kyrgyzstana – Fevral'skii reiting' [Forty well–known politicians of Kyrgyzstan – the February rating], *Slovo Kyrgyzstana*, 11–12 March: 4.

—— (1997a) '40 vedyschikh politikov Kyrgyzstana v Fevrale' [Forty well–known politicians of Kyrgyzstan in February], *ResPublica*, 11–17 March.

—— (1997b) 'Geroi i antigeroi: 40 veduschikh politikov v Mae' [Hero and anti-hero: forty well–known politicians in May], *Slovo Kyrgyzstana*, 5 June.

Isaev, K., Shaidullieva, T. and Madaliev, M. (1998) 'Sotsial'noe nostroenie: ukhushednie? rasteriannost'? prisposoblenie k peremenam? Rost nadezhdy?' [The social mood: Worsening? Dismayed? Accommodating to changes? Growth of Hope?], *ResPublica*, 17–23 November.

Ismailova, S. (1995) 'Sotsiologicheskoe obrazovanie: problemy stanovleniia i perspektivy' [Sociological education: problems of its creation and perspectives], *Kut Bilim*, 31 May: 6.

Ivanov, V. (1988) 'Perestroika i sotsiologicheskaia nauka' [Perestroika and sociological science], *Sotsiologicheskie issledovaniia*, 1: 7–12.

Ivanov, V. and Osipov, G. (1989) 'Traditions and specific features of sociology in the Soviet Union', in Genov, N. (ed.) *National Traditions in Sociology*, London: Sage.

Joshi, P. (1995) *Social Science and Development: Quest for Relevance*, New Delhi: Har–Anand Publications.

Kabyscha, A. (1990) 'Nauchnyi status istoricheskogo materializma i predmet sotsiologii' [The scientific status of historical materialism and the subject of sociology], *Sotsiologicheskie Issledovaniia*, 4: 15–24.

Kakeev, A. (1990) 'Filosofskaia nauka v Kirgizii', *Izvestiia Akademii nauk Kirgizskoi SSR*, 2: 23–40.

Kalb, D. (2002) 'Globalism and postsocialist process', in Hann, C. (ed.) *Postsocialism: Ideals, ideologies and practices in Eurasia*, London and New York: Routledge.

Kandiyoti, D. (2002) 'Postcolonialism compared: potentials and limitations in the Middle East and Central Asia', *International Journal of Middle East Studies*, 34: 279–97.

Karakeev, K. (1974) *Akademiia Nauk Kirgizskoi SSR* [The Academy of Science of the Kirgiz SSR], Frunze: Izdatel'stvo Ilim.

—— (1974) 'Kul'turnoe stroitel'stvo v Kirgizii, 1930–31' [Cultural construction in Kirgiziia, 1930–31], in *Akademiia Nauk Kirgizskoi SSR*, Frunze: Izdatel'stvo Ilim.

Karim Kuzu, G. (2003) 'Problems of secondary school education in Kyrgyzstan', *Central Asia-Caucasus Analyst*, 26 February.

Karklins, R. (1986) *Ethnic Relations in the USSR: The perspective from below*, Boston, MA: Unwin Hyman.

Kecskemeti, P. (1952) *Essays in the Sociology of Knowledge*, London: Routledge and Kegan Paul.

Keen, M. and Mucha, J. (1994) *Eastern Europe in Transition: The impact on sociology*, Westport, CT: Greenwood Press.

Kellner, D. (2002) 'Theorizing globalization', *Sociological Theory*, 20(3): 285–305.

Kerr, S. (1992) 'Debate and controversy in Soviet higher education reform: reinventing a system', in Dunston, J. (ed.) *Soviet Education under Perestroika*, New York: Routledge.

Khalid, A. (1998) *The Politics of Muslim Cultural Reform: Jadidism in Central Asia*, Berkeley, CA: University of California Press.

Kirgizskaia Sovietskaia sotsialisticheskaia Republica – Entsiklopedia [The Kirgiz SSR – Encyclopedia], (1982) Frunze.

Kitching, G. (1994) *Marxism and Science: Analysis of an obsession*, Pennsylvania: Penn State University Press.

Kodin, E. (1996) 'The reform of Russian higher education: we had, we lost, we gained', *International Higher Education*, August.

Koichuev, T. (1988) 'Ideologiia obnovleniia i razvitie obshchestvennyh nayk v Kirgizskoi SSR' [The ideology of renewal and the development of social science in the Kirgiz SSR], *Izvestiia akademii nauk Kirgizskoi SSR*, 2: 5.

Kuhn, T. (1970) *The Structure of Scientific Revolutions*, Chicago, IL: University of Chicago Press.

Kuklick, H. (1980) 'Boundary maintenance in American sociology: limits to academic "professionalization"', *Journal of the History of the Behavioral Sciences*, 16: 201–219.

Kürti, L. (1996) 'Homecoming: affairs of anthropologists in and of Eastern Europe', *Anthropology Today*, 12(3): 11–15.

Kydralieva, Z. (1998) 'Rol' sotsiologii v sisteme obrazovanii na sovremennom etape' [The role of sociology in the system of education at the current stage (Problems in realizing the cadres of the twenty-first century)], materials from an international conference, Bishkek.

Lakatos, I. and Musgrave, A. (eds) (1970) *Criticism and the Growth of Knowledge*, Cambridge: Cambridge University Press.

Lamy, P. (1976) 'The globalization of American sociology: excellence or imperialism?' *American Sociologist*, 11: 104–14.

Latour, B. (1987) *Science in Action*, Milton Keynes: Open University Press.

—— (1999) *Pandora's Hope: Essays on the reality of science studies*, Cambridge, MA: Harvard University Press.

—— (1999a) 'For David Bloor...and beyond: a reply to David Bloor's "Anti-Latour"', *Studies in History and Philosophy of Science*, 30(1): 113–29.

Lemaine, G., Macleod, R., Mulkay, M. and Weingart, P. (eds) (1976) *Perspectives on the Emergence of Scientific Disciplines*, Paris: Maison des Sciences de l'Homme.

Lenin, V. I. (1952 [1908]) 'Materialism and empirico-criticism: critical notes concerning a reactionary philosophy', in *Collected Works of V. I. Lenin*, Moscow: Foreign Languages Publishing House.

'Leninizm i razvitie obschestvennykh nauk v Srednei Azii i Kazakhstane' [Leninism and the development of the social sciences in Central Asia and Kazakhstan], (1970) *Vestnik Akademii nauk USSR*, 3: 3034.

Lewis, R. (1986) 'Science, nonscience and the Cultural Revolution', *Slavic Review*, 45(2): 286–92.

Liber, G. (1993) Review of Simon, G., 'Nationalism and Policy toward the Nationalities in the Soviet Union: from totalitarian dictatorship to post-Stalinist society', *Russian Review*, 52(4): 572–73.

Lincoln, Y. and Guba, E. (2003) 'Paradigmatic controversies, contradictions, and emerging confluences', in Denzin, N. and Lincoln, Y. (eds) *The Landscape of Qualitative Research: Theories and issues*, London: Sage.

Lisovskaia, E. and Karpov, K. (1999) 'New ideologies in postcommunist Russian textbooks', *Comparative Education Review*, 43(4): 522–43.

Liu, M. (2003) 'Detours from utopia on the Silk Road: ethical dilemmas of neoliberal triumphalism', *Central Eurasian Studies Review*, 2(2): 2–10.

Locke, S. (2001) 'Sociology and the public understanding of science: from rationalization to rhetoric', *British Journal of Sociology*, 52: 1–18.

Lokteva, S. (1991) Interview with Adash Toktosunova, 'Bednye, kotorym mnogo nado' [Poor are they who need much], *Vechernyi Bishkek*, 28 May.

Long, N. and Long, A. (eds) (1992) *Battlefields of Knowledge: The interlocking of theory and practice in social research and development*, London: Routledge.

Lubrano, L. (1977) 'The role of Soviet sociologists in the making of social policy', in Remnek, R. (ed.) *Social Scientists and Policy Making in the USSR*, London: Praeger.

—— (1993) 'The hidden structure of Soviet science', *Science, Technology and Human Values*, 18(2): 147–75.

Luk'ianova, T. (1990) 'Sotsiologiia: uglubliat analiz' [Sociology: extending the analysis. Interview with V. V. Chernyshev], *Kommunist Uzbekistana*, 12: 55–64.

McCarthy, E. (1996) *Knowledge as Culture: The new sociology of knowledge*, London: Routledge.

McDaniel, S. (2003) 'Introduction: the currents of sociology internationally – preponderance, diversity and division of labour', *Current Sociology*, 51(6): 593–97.

MacWilliams, B. (2001) 'For Russia's universities, a decade of more freedom and less money', *Chronicle of Higher Education*, 12 December.

Mandel, W. (1969) 'Soviet Marxism and social science', in Simirenko, A. (ed.) *Social Thought in the Soviet Union*, Chicago, IL: Quadrangle Books.

Mannheim, K. (1991 [1936]) *Ideology and Utopia: An introduction to the sociology of knowledge*, New York: Harcourt and Brace.

Marcuse, H. (1964) *One-Dimensional Man: Studies in the ideology of advanced industrial society*, London: Routledge and Kegan Paul.

Marx, K. and Engels, F. (1991) *The German Ideology*, New York: International Publishers.

Matthews, M. and Jones, T. (1978) *Soviet Sociology: A bibliography, 1964–75*, New York: Praeger.

Megoran, N. (2006) 'The bell tolls for another US-based NGO in Uzbekistan', Available online at: www.eurasianet.org (accessed 20 August 2006).

Meja, V. and Stehr, S. (1990) *Knowledge and Politics: The sociology of knowledge dispute*, London: Routledge.

Merton, R. (1996 [1938]) 'Science and the social order', in Sztompka, P. (ed.) *Robert K. Merton on Social Structure and Science*, Chicago, IL: University of Chicago Press.

—— (1996 [1942]) 'The ethos of science', in Sztompka, P. (ed.) *Robert K. Merton on Social Structure and Science*, Chicago, IL: University of Chicago Press.

—— (1996 [1945]) 'Paradigm for the Sociology of Knowledge', in Sztompka, P. (ed.) *Robert K. Merton on Social Structure and Science*, Chicago, IL: University of Chicago Press.

Messer-Davidow, E., Shumway, D. and Sylvan, D. (1996) 'Disciplinary ways of knowing', in Messer-Davidow, E., Shumway, D. and Sylvan, D. (eds) *Knowledges: Historical and critical studies in disciplinarity*, Charlottesville, VA and London: University Press of Virginia.

'Migratsiia: slukhi i real'nost'' [Migration: rumours and reality], (1992) *Delo No.*, 30 January: 3.

Mikosz, D. (2001) Exchange programs having quiet impact in Central Asia', Available online at: www.eurasianet.org (accessed 15 December 2002).

Mills, C. W. (1963) 'The cultural apparatus', in Horowitz, I. L. (ed.) *Power, Politics and People: The collected essays of C. Wright Mills*, London: Oxford University Press.

—— (1963a) 'Methodological consequences of the sociology of knowledge', in Horowitz, I. L. (ed.) *Power, Politics and People: The collected essays of C. Wright Mills*, London: Oxford University Press.

Ministry of Education, Science and Culture of the Kyrgyz Republic (1994) *Gosudarstvennyi obrazovatel'nyi standart bazogo vyshego obrazovaniia po napravleniiu G.12 (521200) 'Sotsiologiia'* [State educational standard for basic higher education in, G.12 (521200) 'Sociology'], Bishkek: Ministry of Education, Science and Culture.

Ministry of Education, Science and Culture of the Kyrgyz Republic (1999) Kvalifikatsionnye trebovaniia k uroven'iu podgotovki bakalvra po napravleniiu G.12 'Sotsiologiia' [Qualifying requirements for the level of preparation for the bachelor's degree in G.12 'Sociology'], signed by T. Ormonbekov (Kyrgyz Ministry of Education) and Dr John Clark (Academic Council of AUK), unpublished document, archive of the Sociology Department, American University–Central Asia.

—— (2000) 'Litsenziia na pravo vedeniia obrazovael'noi deiatel'nosti v sfere professional'nogo obrazovaniia', AL No. 159 [License to carry out educational work in the sphere of professional education, AL No. 159], Bishkek: Ministry of Education, Science and Culture.

—— (2001) *Ob itogakh gosudarstvennoi attestatsii Amerikanskogo Universiteta v Kyrgyzstane* [The results of the state attestation for the American University in Kyrgyzstan], Bishkek: Ministry of Education, Science and Culture.

Ministry of Education, Science and Culture of the Kyrgyz Republic, State Attestation Commission (2002) 'Otchet gosudarstvennoi attestatsionnoi komissii po attestatsii vypusnikov Amerikanskogo Universiteta v Tsentral'noi Azii' [Report of the State Attestation Commission for the attestation of graduates of the American University–Central Asia], unpublished document, archive of the Sociology Department, American University–Central Asia.

Moore, David Chioni (2001) 'Is the postcolonial the post- in post-Soviet? Towards a global postcolonial critique', *PMLA*, 116(1): 111–28.

Mulkay, M. (1976) 'Norms and ideology in science', *Social Science Information*, 15: 637–56.

—— (1979) *Science and the Sociology of Knowledge*, London: Allen and Unwin.

—— (1991) 'Science in Society', in Review of Cozzens, S. and Gieryn, T. (eds) *Contemporary Sociology*, 20(3): 456.

Mulkay, M. and Gilbert, N. (1982) 'Accounting for error: how scientists construct their social world when they account for correct and incorrect belief', *Sociology*, 16: 165–83.

Myrskaia, E. (1991) 'Soviet sociology: fateful history and present-day paradoxes', *Canadian Journal of Sociology*, 16(1): 75–78.

Naby, E. (1993) 'Publishing in Central Asia', *Central Asia Monitor*, 1.

Nesvetailov, G. (1995) 'Centre-periphery relations and the transformation of post-Soviet science', *Knowledge and Policy: The International Journal of Knowledge Transfer and Utilization*, 8(2): 53–67.

Newman, J. (2003) Interview by the author, American University–Central Asia, Bishkek, Kyrgyzstan, 2 May.

Nichols, P. (1997) 'Creating a market along the silk road: a comparison of privatization techniques in Central Asia', *NY University Journal of International Law and Politics*, 29(3): 299–336.

Nove, A. and Newth, J. (1967) *The Soviet Middle East: A model for development?* London: Allen Unwin.

Nurova, S. (2000) 'Vospominaniia o Tabaldieve A.T.', in *A. T. Tabaldiev i sovremennost': materially filosofskix chtenii, posviaschennyx 60-letnemu iuibeleiu doktora…A.T. Tabaldieva* [A.T. Tabaldiev and the contemporary time: material from philosophical readings, dedicated to the 60th jubilee of A. T. Tabaldiev], Bishkek: National Academy of Science of the Kyrgyz Republic, Institute of Philosophy and Law.

—— (2001) 'Sotsiologicheskoe izuchenie mezhetnicheskikh otnoshenii v Kyrgyzstane' [The sociological study of interethnic relations in Kyrgyzstan], *Dialog tsivilizatsii na velikom shelkovom puti: Tsentral'naia Aziia—vchera, segodnia, zavtra.* Materialy mezhdunarodnoi konferentsii [The dialogue of civilizations on the great silk road: Central Asia yesterday, today, tomorrow], material from an international conference, 26–27 September, Bishkek.

—— (2003) Interview by the author, Bishkek Humanitarian University, Bishkek, Kyrgyzstan, 2 April.

Nurova, S. and Shaimergenova, T. (2000) *Sotsiologicheskie issledovaniia v rabote chitatel'skikh zalov i abonementov bibliotek: metod razrabotka* [Sociological research in the work of readers' halls and library borrowers: methods of elaboration], Bishkek.

'Ob uchrezhdenii Akademii nauk Kirgizskoi SSR' [On the establishment of the Academy of Science of the Kirgiz SSR] (1962), 'Sovietskaia Kirgiziia', in *Razvitie nauki v sovetskom Kirgizstane* [The development of science in Soviet Kirgizstan], Frunze.

'Obychnyi prepodavatel' obychnogo vuza' [An ordinary teacher of an ordinary higher educational institution] (2000), 'Skol'ko stoit otuchit'sia?' [How much does it cost to complete an education?], *Argumenty i fakty – Kazakhstan*, 6 November.

Omuraliev, N. (1997) 'Sotsiologicheskii analiz mezhetnicheskikh otnoshenii v Kyrgyzstane' [A sociological analysis of interethnic relations in Kyrgyzstan], 'Etnicheskii mir' [The ethnic world: a bulletin of information and analysis]. Available online at: www.assamblea.kg/em_engl.html# (accessed 12 September 2003).

—— (2003) Interview by the author, Centre for Sociological Research, National Academy of Science of the Kyrgyz Republic, Bishkek, Kyrgyzstan, 4 July.

Omurkulova, C. (2003) Interview by the author, International Board for Research and Exchange (IREX), Bishkek, Kyrgyzstan, 14 February.

Oruzbaeva, B. (ed.) (1980) *Kyrgyz Soviet Entsiklopediasy* [Kyrgyz Soviet Encyclopaedia], Frunze.

Osmonalieva, R. (1994) 'Otkuda vy cherpaete svoi poznaniia o privatizatsii?' [From where do you get your knowledge about privatization?], *Svobodnye gory*, 29 July: 7.

—— (1995) Interview with Kusein Isaev, 'Nauka – glavnaia sila, na kotoroi rastsvetaet obschestvo' [Science is the main force upon which society flourishes], *Kut Bilim*, 18 January: 4.

Osmonov, O. (2001) *Aktual'nye problemy gumanitarnykh i sotsial'nykh nauk Kyrgyzstana: sbornik nauchnykh statei* [Pressing problems of the humanitarian and social sciences of Kyrgyzstan: a collection of scientific articles], Bishkek.

Osorov, Z. (2002) 'Higher education is a big business', *The Times of Central Asia*, 18 October.

Outhwaite, W. and Ray, L. (2005) *Social Theory and Postcommunism*, Oxford: Blackwell Publishing.

Panarin, S. (1994) 'The ethnohistorical dynamics of Muslim societies within Russia and the CIS', in Mesbahi, M. (ed.) *Central Asia and the Caucasus after the Soviet Union: Domestic and international dynamics*, Gainesville, FL: University Press of Florida.

Park, A. (1976) 'Soviet policy in Soviet Asia', in Dodge, N. (ed.) *The Soviets in Asia*. Proceedings of a symposium sponsored by the Washington Chapter of the American Association for the Advancement of Slavic Studies and the Institute for Sino–Soviet Studies, George Washington University, 19–20 May.

Peet, R. and Watts, M. (1993) 'Introduction: development theory and environment in an age of market triumphalism', *Economic Geography*, 69(3): 227–53.

Pertierra, R. (1997) *Explorations in Social Theory and Philippine Ethnography*, Diliman, Quezon City: University of the Philippines Press.

Phipps, R. and Wolanin, T. (2001) 'Higher education reform initiatives in Kyrgyzstan: an overview', initial report to the Ministry of Education and Culture, Kyrgyz Republic, produced through the support of the Eurasia Foundation.

Pine, P. and Bridger, F. (1997) *Surviving Post-Socialism: Local strategies and regional responses in Eastern Europe and the former Soviet Union*, London: Routledge.

Popkewitz, T. (1991) *A Political Sociology of Educational Reform: Power/Knowledge in teaching, teacher education, and research*, New York: Columbia University Press.

Popovsky, M. (1979) *Manipulated Science: The crisis of science and scientists in the Soviet Union today*, New York: Doubleday.

Potter, J. (1996) *Representing Reality: Discourse, rhetoric and social construction*, London: Sage.

Prashad, V. (1998) Review of Chomsky, N. (ed.) 'The Cold War and the University', in *Frontline: India's national magazine*, 15(23). Available online at: www.hinduonnet.com/fline/fl1532/15320720.htm (accessed 24 March 2002).

Radio Free Europe (2000) 'Association of sociologists and political scientists marks fifth anniversary', *Kazakh report*. Available online at: www.rferl.org/reports/kazakh-report/2000/09/0–220900.asp (accessed 22 September).

—— (2001) *Kyrgyz News Bulletin*. Available online at: www.eurasianet.org/resource/kyrgyzstan/hypermail/200103/0051.html (accessed 23 March).

Rahman, A. (1983) *Intellectual Colonisation: Science and technology in West–East relations*, New Delhi: Vikas Publishing House.

Raiymbekova, K. (1999) 'Vysshee obrazovanie suverennogo Kyrgyzstana' [Higher education in sovereign Kyrgyzstan], candidate of science dissertation, Kyrgyz State National University.

Rakowska-Harmstone, T. (1972) 'Recent trends in Soviet nationality policy', in Dodge, N. (ed.) *The Soviets in Asia*. Proceedings of a symposium sponsored by the Washington Chapter of the American Association for the Advancement of Slavic Studies and the Institute for Sino-Soviet Studies, George Washington University, 19–20 May.

Ram, U. (1991) 'The Israeli sociological imagination', PhD dissertation, New York: The New School for Social Research.

Razguliaev, Iu. (1995) 'O prezidente: ili khorosho, ili nichego' [About the president: either good or nothing], *Pravda*, 18 March.

Reeves, M. (2002) 'Cultivating citizens of a "new type": the politics and practice of educational reform at the American University in Kyrgyzstan', unpublished paper, Department of Social Anthropology, University of Cambridge.

—— (2003) 'The role of universities in the transformation of societies: case study of the American University–Central Asia', report conducted for the Centre of Higher Education Research, information of the Open University of Great Britain and the Association of Commonwealth Universities, with support from the Open Society Institute.

—— (2003a) 'The theoretical and ethical issues raised by collaborative research projects involving individuals or institutions embedded in one's context of study', unpublished paper, Department of Social Anthropology, Cambridge University.

—— (2004) 'Academic integrity and its limits in Kyrgyzstan', *International Higher Education*, 37.

Remington, R. (1988) *The Truth of Authority: Ideology and communication in the Soviet Union*, Pittsburgh, PA: University of Pittsburgh Press.

Restivo, S. and Loughlin, J. (2000) 'The invention of science', *Cultural Dynamics*, 12(2): 135–59.

Reynolds, L. and Reynolds, J. (1970) *The Sociology of Sociology: Analysis and criticism of the thought, research and ethical folkways of sociology and its practitioners*, New York: David McKay and Co.

Robinson, W. (1996) 'Globalization, the world-system and "democracy promotion" in US foreign policy', *Theory and Society*, 25(5): 615–65.

Ro'i, J. (1995) 'The secularization of Islam and the USSR's Muslim areas', in Ro'i, J. (ed.) *Muslim Eurasia: Conflicting legacies*, London: Frank Cass.

Rothman, R. (1972) 'A dissenting view on the scientific ethos', *British Journal of Sociology*, 23(1): 102–8.

Ruble, B. (1999) Foreward to the Humanities and Social Sciences in the Former Soviet Union: an assessment of need, unpublished report prepared by the Kennan Institute, Woodrow Wilson Center for the Carnegie Corporation of New York and the MacArthur Foundation.

Ruble, B. and Skvortsov, L. (eds) (1993) *A Scholars' Guide to Humanities and Social Sciences in the Soviet Successor Sates*, Armonk, New York: M.E. Sharpe.

Rumer, B. (2000) *Central Asia and the New Global Economy*, London and New York: M.E. Sharpe.

Rumiantsev, A. and Osipov, G. (1968) 'Marksistskaia sotsiologiia i konkretnye sotsial'nye issledovaniia' [Marxist sociology and concrete social research], *Voprosy filosofii*, 6: 6–13.

Ryskulov, T. (1998) Interview with K. Isaev, 'Sotsiologiia Kyrgyzstana v mire i u sebia' [Sociology of Kyrgyzstan: in the world and for itself], *Nasha Gazeta*, 26 September: 13.

Ryskulueva, F. (2003) Interview by the author, Ministry of Education and Culture of the Kyrgyz Republic, Bishkek, Kyrgyzstan, 7 July.

Sabloff, P. (ed.) (1999) *Higher Education in the Post-Communist World: Case studies of eight universities*, New York: Garland Publishing.

Sagynbaeva, A. (2000) 'Biznesa bez analiza byt' ne mozhet . . . .' [Business cannot exist without analysis. . . .], *Biznesmen Kyrgyzstana*, 9(41), March: 1.

—— (2003) Interview by the author, SIAR Bishkek, Bishkek, Kyrgyzstan, 22 May.

Sahni, K. (1997) *Crucifying the Orient: Russian orientalism and the colonization of caucasus and central Asia*, Oslo: White Orchid Press.

Said, E. (1978) *Orientalism*, New York: Penguin Books.

Sakwa, R. (1999) *Postcommunism*, Buckingham: Open University Press.

Salehi-Esfahani, S. and Thornton, J. (1998) 'The dilemma of reforming economics education in the post-socialist economy of Uzbekistan: has anything changed?', *Central Asian Survey*, 17(2): 253–65.

Sanghera, B. (2003) Interview by the author, London, 12 September.

Schott, T. (1992) 'Soviet science in the scientific world system: was it autarchic, self-reliant, distinctive, isolated, peripheral, central?', *Knowledge: Creation, diffusion, utilization*, 14(4): 410–39.

—— (1993) 'World science: globalization of institutions and participation', *Science, Technology and Human Values*, 18(2): 196–208.

—— (1998) 'Ties between center and periphery in the scientific world-system: accumulation of rewards, dominance and self-reliance in the center', *Journal of World Systems Research*, 4(2): 112–44.

179

Segerstråle, U. (ed.) (2000) *Beyond Science Wars: The missing discourse about science and society* (Suny Series: Science, Technology and Society), Albany, NY: State University of New York Press.

Shahrani, M. (1994) 'Muslim Central Asia: Soviet development legacies and future challenges', in Mesbahi, M. (ed.) *Central Asia and the Caucasus after the Soviet Union: Domestic and international dynamics*, Florida: University Press of Florida.

Shaidullaeva, T. (1992) 'The structure and function of the modern Kyrgyz rural family'. Thesis brief (*avtoreferat*) for candidate dissertation in Social Structure, Social Institutes and Ways of Life. Department of Sociology and Psychology, Kyrgyz Women's Pedagogical Institute, Bishkek, Kyrgyzstan.

Shalin, D. (1978) 'The development of Soviet sociology, 1956–76', *Annual Review of Sociology*, 4: 171–91.

—— (1990) 'Sociology for the glasnost era: institutional and substantive changes in recent Soviet sociology', *Social Forces*, 68(4): 1019–39.

Shanin, T. (1986) 'Theories of ethnicity: the case of a missing term', *New Left Review*, 1(158).

Sharshekeeva, K. (2001) 'Amerikanskii universitet v Kyrgyzstane kak odna is sovremennykh modelei razvitiia vysshei shkoly' [The American University in Kyrgyzstan as a contemporary model for the development of higher education institutions], candidate of science dissertation, American University in Kyrgyzstan.

Sheehan, H. (1993) *Marxism and the Philosophy of Science*, Atlantic Highlands, NJ: Humanities Press International.

Sherstobitov, V. (1987) 'Vysokoe prednaznachenie obshchestvennykh nauk' [The highest mission of the social sciences], *Izvestiia akademii nauk Kirgizskoi SSR, Obshchestvennye nauki*, 1: 3–8.

Shils, E. (1970) 'Tradition, ecology and institution in the history of sociology', *Daedalus*, 99(4): 760–825.

—— (1988) 'Center and periphery: an idea and its career, 1935–87', in Greenfeld, L. and Martin, M. (eds) *Center: Ideas and institutions*, Chicago, IL: University of Chicago Press.

Shlapentokh, V. (1987) *The Politics of Sociology in the Soviet Union*, Boulder, CO: Westview Press.

Shumway, D. and Messer-Davidow, E. (1991) 'Disciplinarity: an introduction', *Poetics Today*, 12(2): 201225.

Simirenko, A. (1966) *Soviet Sociology: Historical antecedents and current appraisals*, Chicago, IL: Quadrangle Books.

—— (1969) 'International contributions by Soviet sociologists', in Simirenko, A., *Social Thought in the Soviet Union*, Chicago, IL: Quadrangle Books.

—— (1969a) 'The development of Soviet social science', in Simirenko, A., *Social Thought in the Soviet Union*, Chicago, IL: Quadrangle Books.

Sklair, L. (1995) *The Sociology of the Global System*, 2nd edn, New York: Prentice Hall.

Skorodumova, E. (1998) 'Ot granta do tiurmy – vsego odin shag?' [Only one step from a grant to jail?] *Vechernyi Bishkek*, 27 November: 7.

Skripkina, S. (1983) 'K voprosu o povyshenii kvalifikatsii nauchnykh kadrov obschestvovedov Kirgizii' [Toward the question of raising the qualifications of scientific cadres of social scientists in Kirgizia], *Sveriaia po Marksu i Leninu shag*, Frunze.

Small, M. (1999) 'Departmental conditions and the emergence of new disciplines: two case studies in the legitimation of African-American studies', *Theory and Society*, 28: 659–70.

Smanbaev, A. (1986) 'Pomiat' ob uchenom' [Memorial about a teacher], *Uchitel' Kirgizstana*, 26 December: 2.

Smart, B. (1994) 'Sociology, globalization and postmodernity: comments on the "Sociology for One World" thesis', *International Sociology*, 9(2): 149–59.

Smith, D. (1988) *The Everyday World as Problematic: A feminist sociology*, Milton Keynes: Open University Press.

Smith, G., Law, V., Wilson, A., Bohr, A. and Allworth, E. (1998) *Nation-Building in the Post-Soviet Borderlands: The politics of national identity*, Cambridge: Cambridge Universtiy Press.

Sorokina, Z. (1989) 'Sovetuiut sotsiologi' [Sociologists advise], *Sovietskaia Kirgiziia*, 10 November.

Sotsiologicheskoe obschestvo Kyrgyzstana (n.d.) ' "Igry" pravitel'stva i PROON vokrug sotsiologii: protest Sotsiologicheskogo obschestva Kyrgyzstana' [The 'games' of the government and UNDP around sociology: a protest by the Sociological Society of Kyrgyzstan], unpublished document, Bishkek.

Sotsiologicheskoe obschestvennoe ob'edinenie (1999) 'Ustav sotsiologicheskogo obschestvennogo ob'edineniia (SOO)' [Constitution of the union of sociologists], unpublished document, Bishkek.

Spector, R. (2004) 'The transformation of Askar Akaev, President of Kyrgyzstan', paper 2004–2, Institute of Slavic, East European and Eurasian Studies, Berkeley Programme in Soviet and Post-Soviet Studies, University of California, Berkeley. Available online at: repositories.cdlib.org/iseees/bps/2004_2–spec (accessed 25 August 2006).

Sutherland, K. (1992) 'Perestroika in the Soviet general school: from innovation to independence?', in Dunston, J. (ed.) *Soviet Education under Perestroika*, New York: Routledge.

Swartz, D. (1997) *Culture and Power: The sociology of Pierre Bourdieu*, Chicago, IL: University of Chicago Press.

Swidler, A. and Arditi, J. (1994) 'The new sociology of knowledge', *Annual Review of Sociology*, 20: 305–29.

Swiss Agency for Development and Cooperation (2000) 'Key actors of Kyrgyzstan', Bern.

Sydykova, Z. (1998) Interview with Kusein Isaev, 'U sotsiologii trudnoe proshloe, a buduschee takoe zhe neopredelennoe, kak i nastoiaschee' [Sociology has had a difficult past and its future, like its present, is equally as uncertain], *ResPublica*, 25 February.

Tabyshaliev, S. T. (1984) 'Razvitie obschestvennykh nauk v Kirgizstane' [The development of the social sciences in Kirgizstan] in *Velikii Oktiabr' i obrazovanii SSSR v sovremennoi ideologicheskoi bor'be* [Great October and USSR education in the contemporary ideological struggle], Frunze.

Tabyshalieva, A. (1983) 'K voprosu o povyshenii kvalifikatsii nauchnykh kadrov obschestvovedov Kirgizii' [Toward the question of raising qualifications of the scientific cadres of social scientists in Kirgizia], *Sveriaia po Marksu i Leninu shag*, Frunze.

—— (1984) 'Krepit' nauchnye sviazi' [To strengthen scientific networks], *Uchitel' Kirgiztstana*, 10 August: 2.

—— (1986) 'K voprosam prakticheskogo ispol'zovaniia obschsetvovedchestkikh issledovanii' [Toward questions of the practical use of social scientific research], *Materialy VII mezhResPublicanskoi nauchnoi konferentsii molodykh uchenykh* [Material from the 7th inter-republican scientific conference of young scholars], Frunze: Ilim.

Tadevosian, E. V. (1963) 'The further convergence of the socialist nations of the USSR', *Voprosy filosofii*, 4.

Taylor, C. (1996) *Defining Science: A rhetoric of demarcation*, Madison, WI: University of Wisconsin Press.

Tchoroev, T. (2002) 'Historiography of post–Soviet Kyrgyzstan', *International Journal of Middle East Studies*, 34: 351–74.

Thompson Klein, J. (1996) 'Blurring, cracking, and crossing: permeation and the fracturing of discipline', in Messer-Davidow, E., Shumway, D. and Sylvan, D. (eds) *Knowledges: Historical and critical studies in disciplinarity*, Charlottesville, VA and London: University Press of Virginia.

Tillett, L. (1964) 'Soviet second thoughts on Tsarist colonialism', *Foreign Affairs*, 42(2): 309–19.

Tishin, A. (1980) 'Dostizheniia sotsiologii – VUZam' [The achievements of sociology into higher education institutions], *Mugalimder*, 31 October: 3.

—— (1988) 'Opyt sotsiologii – proizvodstvu' [The experience of sociology into production], *Kommunist Kirgizstana*, 1: 62–67.

Tishin, A. (1989) *Chitateli i gazeta v zerkale sotsiologii: materialy issledovaniia raionnoi pechati Kirgizii* [Readers and newspapers in the mirror of sociology: material from research on the regional press of Kirgizia], Frunze.

—— (1998) 'Razvitie sotsiologii v Kyrgyzkoi Respublike' [The development of sociology in the Kyrgyz Republic] in A.I. Tishin, *Kurs kratkix leksii po sotsiologii* [A course of short lectures on sociology], Bishkek, Center for Strategic Research, State Service, and Political Science, Kyrgyz State National University.

—— (2000) 'Slovo ob uchenom' [A word about a scholar] in *A. T. Tabaldiev i sovremennost': materially filosofskix chtenii, posviaschennyx 60-letnemu iuibeleiu doktora...A.T. Tabaldieva* [A.T. Tabaldiev and the contemporary time: material from philosophical readings, dedicated to the 60th jubilee of A.T. Tabaldiev], Bishkek: National Academy of Science of the Kyrgyz Republic, Institute of Philosophy and Law.

—— (2003) Interview by the author, Kyrgyz National University, Bishkek, Kyrgyzstan, 18 March.

Tishin, A., Bekturganov, K. and Shakitov, S. (1998) 'Sotsiologiia lzhi ili lozh' sotsiolog?' [A sociology of lies or the lie of the sociologist?'], *Slovo Kyrgyzstana*, 2 April: 4.

Tishin, A., Svitich, L., Tarasov, A. and Akulov, F. (1989) *Sotsiologicheskoe zerkalo rainnoi pechati: progr. i instrumentarii issled., provedennogo v Kirgizii* [A sociological reflection of the regional press: programme and instrumentation of research conducted in Kirgizia], Frunze: Goskomizdat Kirgiz SSR.

Tishkov, V. (1998) 'US and Russian anthropology: unequal dialogue in a time of transition', *Current Anthropology*, 39(1): 1–17.

Toktosunova, A. and Sukhanova, L. (1990) 'Perekhod na samoupravlenie' [The transition to self-governance], *Sovietskaia Kirgiziia*, 4 August.

Tomlinson, J. (1991) *Cultural Imperialism*, Baltimore, MD: Johns Hopkins University Press.

Torres, C. (1999) 'Critical theory and political sociology of education: arguments', in Popkewitz, T. (ed.) *Critical Theories in Education: Changing terrains of knowledge and politics*, London: Routledge.

Toshchenko, Z. (1998) 'Kazakhstan: stanovlenie natsional'noi sotsiologii' [Kazakhstan: the creation of a national sociology], *Sotsiologicheskie issledovaniia*, 3: 3–4.

Turner, S. and Turner, J. (1990) *The Impossible Science*, CA: Sage Publications.

Urban, P. and Lebed, A. (1971) *Soviet Sciences, 1917–1970*, Part I: Academy of Science of the USSR, Metuchen, NJ: Scarecrow Press.

Vasquez, J. (1995) 'The post-positivist debate: reconstructing scientific enquiry and international relations theory after Enlightenment's fall', in Booth, K. and Smith, S. (eds) *International Relations Theory Today*, Cambridge: Polity Press.

Verdery, K. (2002) 'Whither postsocialism?', in *Postsocialism: Ideals, ideologies and practices in Eurasia*, London and New York: Routledge.

Vlasova, T. (1989) 'Sluzhba sotsial'nogo razvitiia – opyt, problemy' [The service of social development – experience, problems], *Kommunist Kirgizstana*, 1(January): 41–44.

Vucinich, A. (1974) 'Marx and Parsons in sociology', *Russian Review*, 33(1): 1–19.

Wagner, P. and Wittrock, B. (1990) 'Analyzing social science: on the possibility of a sociology of the social sciences', in Wagner, P., Wittrock, B. and Whitley R. (eds) *Discourses on Society: The shaping of the social science disciplines*, Boston, MA and London: Kluwer.

Walker, R. (1989) 'Marxism–Leninism as discourse: the politics of the empty signifier and the double-bind', *British Journal of Political Science*, 19(2): 161–89.

Watts, M. (1993) 'Development I: power, knowledge, discursive practice', *Progress in Human Geography*, 17(2): 257–72.

Weinberg, E. (1974) *The Development of Sociology in the Soviet Union*, London: Routledge and Kegan Paul.

—— (1992) 'Perestroika and Soviet sociology', *British Journal of Sociology*, 43(1): 1–10.

—— (1994) 'Reality and research: current issues in Soviet and Russian studies', *British Journal of Sociology*, 45(1): 133–41.

—— (2004) *Sociology in the Soviet Union and Beyond: Social enquiry and social change*, Aldershot: Ashgate.

Wheeler, G. (1966) *The Peoples of Soviet Central Asia: A background book*, London: The Bodely Head.

Whyte, W. F. (1969) 'The role of the US professor in developing countries', *The American Sociologist*, February: 19–28.

Wirth, L. ([1936] 1991) Preface to Mannheim's *Ideology and Utopia: An introduction to the sociology of knowledge*, New York: Harcourt and Brace.

Wittrock, B., Wagner, P. and Wollmann, H. (eds) (1991) *Social Sciences and Modern States: National experiences and theoretical crossroads*, Cambridge: Cambridge University Press.

Yanowitch, M. (1989) 'The influence of Tat'iana Zaslavskaia on Soviet social thought', in Yanowich, M. (ed.) *A Voice of Reform: Essays by Tat'iana Zaslavskaia*, New York: M.E. Sharpe.

Yurchak, A. (1997) 'The cynical realism of late socialism: power, pretense and the *Anekdot*', *Public Culture*, 9: 166–88.

—— (2003) 'Soviet hegemony of form: everything was forever, until it was no more', *Comparative Studies in Society and History*, 45(3): 480–510.

Zanca, R. (2000) 'Intruder in Uzbekistan: walking the line between community needs and anthropological desiderata', in De Soto, H. and Dudwick, N. (eds) *Fieldwork Dilemmas: Anthropologists in postsocialist states*, Madison, WI: University of Wisconsin Press.

Zarlikbekov, J. (1998) Interview with Kusein Isaev, 'Sotsiologiany saiasii soiku kylgylary bar' [They want to make a political prostitute out of sociology], *Asaba*, 6 March.

Zaslavskaia, T. (1989) 'The role of sociology in accelerating the development of Soviet society', in Yanowitch, M. (ed.) *A Voice of Reform: Essays by Tat'iana Zaslavskaia*, New York: M.E. Sharpe.

Zaslavsky, V. (1977) 'Sociology in the contemporary Soviet Union', *Social Forces*, 44(2): 330–53.

Zborovskii, G. E. (2001) 'Suschestvuet li regional'naia sotsiologiia?' [Does regional sociology exist?], *Sotsiologicheskie issledovaniia*, 1: 101–08.

Zestov, I. (1985) *Soviet Sociology: A study of lost illusions in Russia under Soviet control of society*, Fairfax, VA: Hero Books.

Zhivogliadov, V. (1990) 'Na shto nam nadeiat'sia?' [What can we hope for?], *Sovietskaia Kirgiziia*, 24 February.

Zhorobekova, E., Kunin, A. and Zhusibaliev, A. (1995) 'Lideri Kyrgyzstana: mnenie potentsial'nykh izbiratelei po Oshskomu regionu' [Leaders of Kyrgyzstan: the opinion of potential voters in the Osh region], *Svobodnye gory*, 2 December.

INDEX

development 7, 94, 119, 145; institutions 95; socialist 74, 146
disciplinarity 3, 5, 35, 47, 63, 73, 75, 93–94, 99, 102–04, 117–18, 144

Eastern Europe 88
epistemology x, 4, 82, 92–93, 100, 103, 116, 124, 134, 138, 148; and 'epistemic negotiation' xii; and late capitalism 28, 98; positivist 7, 16, 140, 146; *see also* knowledge, legitimization of
ethnicity 96
ethnic relations 78; sociology of 57, 59, 61, 70, 144
Eurasia Foundation 85

Fanon, F. 32, 77
Foucault, M. 10, 21, 22
Frunze Polytechnic Institute 65, 71, 77, 79, 95
functionalism 83

gender 96–97
generation 96, 111, 118
globalization: and knowledge 2, 28, 123; and post-socialism xi, 149
goszakazchiki see *zakazchiki*

hegemony, intellectual 36, 64, 101
higher education: commercialization of 87, 98, 106, 142; funding of 82, 84, 91, 95, 101, 109–10, 118; internationalization of 108; markets of 86, 95, 97; mobility in 101; reform 93
Historical Materialism, departments of 51

ideology critique, and sociology of knowledge 19
indigenous sociology *see* sociology, national
industrial sociology *see* sociology, industrial
institutions local 90–91
intelligentsia 77, 88
internationalization *see* higher education, internationalization of; sociology, internationalization of
International Labor Organization 85
International Monetary Fund 85

international organizations 85, 91, 97, 106, 108, 110, 118; *see also by name and* development, institutions
Isaev, K. 13, 16, 24, 47, 65, 70, 77, 95, 99, 130, 139

Jadids 37

Kazakhstan 87, 88, 90
Kirgiz State University *see* Kyrgyz National University
knowledge: autonomy of 4, 16, 23, 32, 78, 84–85, 88, 92, 129, 145, 147; and colonialism 9, 30, 32, 33, 38; critical theories of xiii; democratization of 136, 145; 'expert' vs. 'lay' 137; Islamic 38, 39; legitimization of 100, 103, 113, 121, 125, 128–29, 139; philosophy of 14; politics of ix, 3, 6, 10, 13, 17, 21, 59, 86–87, 92; and power 2, 7, 15, 55, 62, 75, 87, 90, 92, 103, 129, 131, 145–46, 148; relevance of 87, 121, 133–34, 141, 146; scientific xiii, 16, 20; sociology of (*see* sociology of knowledge); as technology 55; *see also* globalization, and knowledge; sociology, and politics
konfliktologiia 83
korenizatsiia 38
Kyrgyz National University 47, 89, 107
Kyrgyzstan: national independence 4, 77, 81, 85, 87; relationship to Russia 44, 56–57, 58, 70, 82, 84, 88, 99; sociology in 14, 26, 40, 43, 47–48, 52, 54, 61, 82, 94

labour market 101–02
'late socialism' 84, 98, 126, 145

MacArthur Foundation 85, 95
Mannheim, K. 1, 19, 20, 27, 47
market: discourses of 63, 76; logic of 4, 26, 108; research 128
marketization *see* higher education, commercialization of
Marxism x, 16, 19, 41, 82, 96, 146; and positivism 17, 146
Marxism–Leninism 4, 5, 45, 94; and 'national relations' 58, 78; and philosophy of science 104, 125, 141; and sociology 7, 42, 50, 52, 55, 66, 82, 86, 127, 146

Merton, R. 19, 83, 141
modernity 29; and Orientalism 45;
projects of in Central Asia 33, 37–38,
125, 148; and sociology 49, 63, 67,
123, 134

national: 'relations' (*see* ethnic relations);
sociology (*see* sociology, national)
'nationalities' 54
neo-colonialism xiv, 32, 85, 92
neo-liberal: development x, 29;
hegemony 2, 9
non-governmental organizations 119;
see also international organizations
Nurova, S. 60

Objectivity: in research methodology x,
xii, 27, 131, 138, 140; in social science
6, 16, 21, 77, 93, 124, 135, 145
Occidentalism xi, 15, 35, 44–46, 112, 116
open society 83
Open Society Institute ix
Orientalism xi, 1, 32, 34–35 *passim*,
44–46, 112, 146
otechestvennaia nauka 88

perestroika 47, 57, 66; and sociology 66,
75, 88, 92, 98, 126
post-colonialism x, xi, 9, 119, 122; and
knowledge 32, 110; and post-Soviet xi,
29, 30, 83, 106; and social science 3, 90
post-Soviet 2, 23; academic culture ix,
xi, xii, 85; fieldwork xi; higher
education in 82; and post-colonialism
xi, 30, 110; science 4; social science 7,
16, 81, 147; sociology 25;
universities 48; *see also*
post-colonialism, and post-Soviet
power: logic of 46, 145, 147; in research
relationships x, xi, xii; *see also*
knowledge, and power; sociology,
and politics
private higher educational institutions
see Universities, private
privatization 76, 129
'pseudoscience' *see* science, demarcation
from non-science
public opinion definitions of 125–26;
research on 55, 68, 75,
125–29, 130–31
public social science *see* social science,
public

ratings, political: research on 133–34
research: funding 130, 143; methodology
125, 127, 131, 133, 135, 136–37, 140;
and ownership of data 85–86, 96, 110,
119; relationship to teaching 104, 121;
see also knowledge, politics of; public
opinion, research on
ResPublica 124
Russian Sociological Society 41

Sagynbaeva, A. 107
science: authority of 98, 129, 143;
demarcation from non-science 3, 91,
128–29, 131, 136, 140, 143; devaluation
in post-Soviet society 48; ethos of 4,
124, 135, 141–44; global system of 7,
28, 36, 90; ideologies of 33, 77, 94, 104,
124, 145; logic of 4, 93; rhetoric of 18,
34, 89, 93, 124, 129, 131, 135, 143, 147;
sociology of (*see* sociology of science);
see also Soviet Union, science in
Scientific Communism 66, 69;
departments of 47, 71
scientific management 55–56, 74, 125
scientific politics 3, 87, 104, 132
Slovo Kyrgyzstana 124
social policy 95
social science: funding of 84–85; politics
of 4, 6, 16, 30–32, 34; public 123–24;
reform of 18, 68, 144, 145
sociology: applied 95, 98–99;
'bourgeois' 41, 50, 53, 57, 63, 142;
commercialization of 85–86, 99, 116,
142; 'critical' xiv, 50; definitions of
48–50, 62, 65, 89, 91, 94, 98, 102–04,
112–14, 120–21, 123, 132, 144, 146;
empirical 50, 104; funding of (*see*
higher education, funding of); industrial
53, 55–57, 127; institutionalization of 2,
3, 14, 15, 47, 73, 84, 93, 99, 127, 141,
146; internationalization of 83, 90, 112;
laboratories 47, 51, 60–64 *passim*, 89,
95, 118, 144; and media 123, 128, 135;
national 80–81, 87–92 *passim*, 114,
127–28; national standards in Kyrgyzstan
99–101 *passim*; and philosophy 53; and
politics 103, 106, 123, 129, 131–33,
135, 139–40, 142, 147; public (*see* social
science, public); relationship to state
119, 133–34; role of, in Central Asia 34,
53, 55, 56–57, 69, 73, 89, 99, 101, 104,
119, 121, 124, 126, 130, 146, 149–50;